Unification of the Seven Boson Interactions based on the Riemann-Christoffel Curvature Tensor

Modified Galactic Scale Gravity
Explicit Quark Confinement
Interactions between Interactions

STEPHEN BLAHA
BLAHA RESEARCH

Pingree Hill Publishing

ISBN: 978-0-9984575-0-5

Cover Credits

Rev. 00/00/01 December 12, 2016

Some Other Books by Stephen Blaha

All the Megaverse! Starships Exploring the Endless Universes of the Cosmos using the Baryonic Force (Blaha Research, Auburn, NH, 2014)

SuperCivilizations: Civilizations as Superorganisms (McMann-Fisher Publishing, Auburn, NH, 2010)

Universes and Megaverses: From a New Standard Model to a Physical Megaverse; The Big Bang; Our Sister Universe's Wormhole; Origin of the Cosmological Constant, Spatial Asymmetry of the Universe, and its Web of Galaxies; A Baryonic Field between Universes and Particles; Flatverse Extended Wheeler-DeWitt Equation (Blaha Research, Auburn, NH, 2014)

PHYSICS IS LOGIC PAINTED ON THE VOID: Origin of Bare Masses and The Standard Model in Logic, U(4) Origin of the Generations, Normal and Dark Baryonic Forces, Dark Matter, Dark Energy, The Big Bang, Complex General Relativity, A Megaverse of Universe Particles (Blaha Research, Auburn, NH, 2015).

PHYSICS IS LOGIC Part II: The Theory of Everything, The Megaverse Theory of Everything, U(4)⊗U(4) Grand Unified Theory (GUT), Inertial Mass = Gravitational Mass, Unified Extended Standard Model and a New Complex General Relativity with Higgs Particles, Generation Group Higgs Particles (Blaha Research, Auburn, NH, 2015).

The Origin of Higgs ("God") Particles and the Higgs Mechanism: Physics is Logic III, Beyond Higgs – A Revamped Theory With a Local Arrow of Time, The Theory of Everything Enhanced, Why Inertial Frames are Special, Universes of the Mind (Blaha Research, Auburn, NH, 2015).

The Origin of the Eight Coupling Constants of The Theory of Everything: U(8) Grand Unified Theory of Everything (GUTE), S^8 Coupling Constant Symmetry, Space-Time Dependent Coupling Constants, Big Bang Vacuum Coupling Constants, Physics is Logic IV (Blaha Research, Auburn, NH, 2015).

New Types of Dark Matter, Big Bang Equipartition, and A New U(4) Symmetry in the Theory of Everything: Equipartition Principle for Fermions, Matter is 83.33% Dark, Penetrating the Veil of the Big Bang, Explicit QFT Quark Confinement and Charmonium, Physics is Logic V (Blaha Research, Auburn, NH, 2015).

The Periodic Table of the 192 Quarks and Leptons in The Theory of Everything: The U(4) Layer Group, Physics is Logic VI (Blaha Research, Auburn, NH, 2015).

New Boson Quantum Field Theory, Dark Matter Dynamics, Dark Matter Fermion Layer Mixing, Genesis of Higgs Particles, New Layer Higgs Masses, Higgs Coupling Constants, Non-Abelian Higgs Gauge Fields, Physics is Logic VII (Blaha Research, Auburn, NH, 2015)

Unification of the Strong Interactions and Gravitation: Quark Confinement Linked to Modified Short-Distance Gravity; Physics is Logic VIII (Blaha Research, Auburn, NH, 2016).

MoND: Unification of the Strong Interactions and Gravitation II, Quark Confinement Linked to Large-Scale Gravity, Physics is Logic IX (Blaha Research, Auburn, NH, 2016).

CQMechanics: A Unification of Quantum & Classical Mechanics, Quantum/Semi-Classical Entanglement, Quantum/Classical Path Integrals, Quantum/Classical Chaos (Blaha Research, Auburn, NH, 2016).

Available on Amazon.com, Amazon.co.uk, bn.com, and other international web sites as well as at better bookstores (through Ingram Distributors).

Preface

This book begins with a discussion of the types of unification in elementary particle physics and gravitation. It concludes that, at present, the unification of the seven interactions: four known and three 'new' interactions, is properly done using the Riemann-Christoffel curvature tensor since these interactions all involve energy-momentum, and they thus affect the curvature tensor to a greater or lesser degree. We describe the three 'new' interactions in detail: a U(4) Generation group interaction, a U(4) Layer group interaction, and a U(4) gauge field interaction due to the Complex General Relativistic Reality group that we construct. Then the curvature tensor is constructed for all seven interactions including General Relativity. A lagrangian for the boson gauge fields is defined. It leads to higher order derivative field equations using the canonical Euler-Lagrange formalism. We then show that there are interactions between the boson interactions – an extension of the usual Standard Model. We also find the form of the Charmonium potential, and modifications of galactic scale gravity of the form of MoND. We suggest that the proton spin puzzle may be resolvable by a gluon-photon contact term that naturally appears in our formalism.

The present work extends previous studies related to MoND and the quark potential.

CONTENTS

1. Types of Unification of Interactions

In this chapter we consider various forms of the unification of particle interactions. We note that particle interactions appear to always involve spin 1 or spin 2 bosons.[1]

1.1 Symbolic ElectroWeak-Type Unification

One type of interaction unification is simply to additively combine the lagrangian terms for each interaction to form a 'total' interaction. The 'unification' is further enhanced by combining gauge fields of separate interactions through rotations such as the ElectroWeak Weinberg angle rotation.[2] Then one can use a covariant derivative, which takes advantage of the spin 1 gauge field nature of the interaction, to insert interactions in the fermion terms of the total lagrangian.

Gauge field particles appear to obtain a mass through the symmetry breaking Higgs Mechanism. The form of the Higgs sector terms is critical to the success of this form of unification. Unfortunately our knowledge of the nature and form of Higgs particle lagrangian terms awaits experimental determination. Thus, in the ElectroWeak case, in particular, the success of unification remains to be proven experimentally. A hopeful factor iss the proof that ElectroWeak theory is renormalizable.

One wonders, however, to what extent a symbolic theoretical unification of this type is truly unification or merely a symbolic 'gluing' of different interactions.

In the case of ElectroWeak theory, should it prove to be superficial, one must recognize that it played a constructive historical role in increasing our understanding of the Electromagnetic and Weak interactions. Thus ElectroWeak theory was a success from that viewpoint regardless of its future role in Physics.

1.2 Group-Based Unification

It is also possible to envision a unification of interactions within one symmetry group. This type of unification has been a dream of many theorists particularly because renormaization studies of the trend of interactions at high energy suggests the various interaction strengths are becoming similar.

---

[1] This feature of interactions make them particularly conducive for use in covariant derivatives, and perhaps more deeply, for use in defining a Riemann-Christoffel curvature tensor that unites the interactions (as we will see).

[2] Although technically such a roation, in itself, is equivalent to the unrotated form.

If, taking a positive attitude, interaction strengths become similar at high energy, one has to wonder if the disparate group structure seen at today's energy can evolve into a union within one large enveloping group. An enveloping group would necessarily have additional generators that cause an interplay between the different groups seen at low energy.

For example, there might be generators that 'couple' the strong and Weak sectors. These generators, and their associated gauge fields, would have to 'materialize' as particle physics goes to higher energies. There is no evidence at present for the appearance of new interaction terms of this sort. Thus unification within a larger enveloping group is much more than achieving an equality of coupling constants.

1.3 Unification Based on Energy-Momentum Considerations

All gauge boson interactions affect the energy and momentum in a physical system. Consequently they must affect the curvature of space and time to a greater or lesser degree. As a result they must contribute to the total Riemann-Christoffel curvature tensor.

Their common appearance in the form of covariant derivative terms leads to a unification framework that encompasses all seven possible interactions: gravity, the Strong interaction, the Weak interaction, the Electromagnetic interaction, and three interactions that appear to be well justified: a U(4) Generation group interaction that leads to the four fermion generations,[3] a U(4) Layer group interaction that leads to four, as yet unseen, layers of fermions,[4] and a General Relativistic U(4) Reality group that augments gravity at galactic distance scales.[5]

Thus we find seven boson interactions. This book constructs a Riemann-Christoffel tensor embodying all these interactions. From this tensor we construct the boson interaction lagrangian sector and show how modified gravity (similar to MoND), quark confinement, and interactions 'between' the interactions follow.

While one might argue for the ambitious goal espoused in section 1.2 it appears that our approach has the advantage of providing an extension of The Standard Model that is subject to experimental verification NOW.

---

[3] Since only three generations are known a U(3) group is not ruled out. However because of four, apparently conserved, particle numbers such as baryon number, we see theoretical indications that U(4) is the correct group. This symmetry is described in detail later in section 11.3

[4] The U(4) Layer group is based on four partially conserved generation layer numbers. If there are only three generations then the group is U(3). See section 11.4 for details.

[5] This group appears when complex General Relativistic transformations are mapped to real-valued General Relativistic transformations. Its interactions modify the Gravity potential at large distances and may partly account for the MoND phenomena. See section 11.2 and chapter 13.

2. Beyond The Standard Model

Using the Riemann-Christoffel curvature tensor approach to unify the seven boson interactions that appear to exist in Nature, we find that we obtain a curvature tensor with the usual Standard Model interactions plus additional interaction effects:

1. Due to the use of a higher order derivative term we find modifications of gravity at galactic distances are possible offering an alternative to the MoND mechanism.

2. A higher order derivative term gives explicit quark confinement and a quark potential with a similar form to the Charmonium potential found by the Cornell group.

3. We describe three 'new' interactions: a U(4) Generation group interaction, a U(4) Layer group interaction, and a U(4) General Relativistic Reality group interaction. Therefore it increases the number of particle interactions to seven. The resulting symmetry group is SU(3)⊗SU(2)⊗U(1)⊗U(4)⊗U(4)⊗U(4) plus real-valued General Relativity.

4. We find that there are additional interactions 'between' the various boson interactions that may account for some of the experimental discrepancies with the conventional Standard Model. See chapters 9 and 12 for a detailed description of the new boson interactions between boson interactions.

While our discussion is confined to the boson sector of the particle lagrangian, the introduction of the fermion sector is relatively simple through the use of covariant derivatives defined in chapters 4, 12 and 13.

We begin with the unification of the four known interactions. We then introduce three additional interactions that seem to be well justified physically, and form the full Riemann-Christoffel curvature tensor and a lagrangian unification.

3. Why Unite the ElectroMagnetism, the Weak and Strong Interactions, and Gravity?

3.1 The Strong Interaction and Gravitation are the Only Unbroken Symmetries

One suggestive qualitative reason for combining ElectroMagnetism, the Strong Interaction and Gravitation in a unified sector within the Theory of Everything is their uniqueness as unbroken symmetries. The ElectroWeak interaction has a broken symmetry. The U(4) Generation group symmetry and the U(4) Layer group symmetry are also broken. The Strong Interaction and Gravitation then stand out as unbroken symmetries. Further, the symmetry of space-time encourages the view that these unbroken symmetries are associated with space-time geometry – not just in four dimensions but in a wider space that includes 4-dimensional space-time but also contains color confined 'bubbles' of a 32-dimensional complex space, which can be mapped to a 16-dimensional complex-valued space. This 16-dimensional space would be confined to hadrons and the quark-gluon plasmas being created in experiments at CERN and Brookhaven.

3.2 Similarity of the PseudoQuantum Vacuum Expectation Values for Gravity and the Strong Interaction

The Strong interaction and gravity have approximately equal vacuum expectation values generating their coupling constants as shown in eq. 1.43 in Blaha (2015d). For the reader's convenience we reproduce slightly edited parts of Blaha (2015d) and (2016c) in appendix B. The coupling constants are shown to be of order of magnitude unity if the Planck mass is chosen to be the common scale. (See appendix B.)

- The strong interaction coupling constant[6] vacuum expectation value $\Phi_1' = g_S = 1.22$
- The ElectroWeak SU(2) coupling constant vacuum expectation value $\Phi_2' = g_{EW} = 0.619$.
- The Electromagnetic U(1) coupling constant vacuum expectation value $\Phi_E' = 0.303$.
- The ElectroWeak U(1) coupling constant vacuum expectation value $\Phi_3' = g'_{EW} = 0.347$.
- The Dark ElectroWeak SU(2) coupling constant vacuum expectation value $\Phi_4' = g_{EWD}$. (7.30)
- The Dark ElectroWeak U(1) coupling constant vacuum expectation value $\Phi_5' = g'_{EWD}$.
- The Layer Group U(4) coupling vacuum expectation value $\Phi_6 = g_V$.
- The Generation gauge field U(4) coupling constant vacuum expectation value $\Phi_7' = g_G$.
- The gravitational coupling constant vacuum expectation value $\Phi_8' = 1$.

---

[6] Based on the running coupling constant value $\alpha_s\ (M_Z^2) = 0.1193 \pm 0.0016$.

We thus see that in the pseudoquantum formalism for the vacuum expectation values of coupling constants $\Phi_1' = g_S = 1.22$ while the gravitation equivalent $\Phi_8' = 1$ is almost the same.The electromagnetic coupling constant $\Phi_E' = 0.303$ is also comparable, as is the Weak SU(2) coupling constant $\Phi_2' = g_{EW} = 0.619$. *We conclude that uniting the four interactions is not unreasonable.*

4. New Formulation of Complex General Relativity, ElectroMagnetism, and the Weak and Strong Interactions

This chapter describes a unified formalism for Complex General Relativity, ElectroMagnetism, the Weak interaction, and the Strong interaction.[7] It is largely based on a 1976 essay by the author (appendix D), a 1975 paper by the author, and Blaha (2016c) and (2015a) as well as earlier books.

4.1 Vierbein (Tetrad) Form of General Relativity

Weyl[8] first introduced the *vierbein* form of General Relativity in order to accommodate fermion spin within General Relativity. Since then a number of papers[9,10] have appeared on the vierbein formalism. The basic relations of the vierbein formalism are:

1. A vierbein (tetrad) has the form[11] $l^{\mu a}(x)$ where μ is a General Relativistic vector index and a is a Special Relativistic index for a point in a 4-dimensional space-time tangent space to a curved space-time. Both indices range from 0 to 3.

2. The curve space-time metric tensor is defined by

$$g_{\mu\nu} = \eta_{ab} l_\mu{}^a(x) l_\nu{}^b(x) \tag{4.1}$$

where η_{ab} = diag(1, -1, -1, -1) is the flat space-time, tangent space metric.

---

[7] This unified theory is unrelated to Kaluza-Klein theories (and derivatives thereof). The theories attempting to unify gravity and Electromagnetism had a different basis and typically relied on a curled fifth dimension. Our theory differs in origin and in detail. We espouse a 16-dimensional space without curling that constitutes what we have called the Megaverse. Kaluza-Klein theories are described in Kaluza, Theodor, "Zum Unitätsproblem in der Physik". Sitzungsber. Preuss. Akad. Wiss. *Berlin. (Math. Phys.)*: 966–972 (1921); Klein, Oskar, "Quantentheorie und fünfdimensionale Relativitätstheorie". Zeitschrift für Physik A **37** (12): 895–906 (1926; Klein, Oskar, "The Atomicity of Electricity as a Quantum Theory Law". *Nature* **118**: 516 (1926); and subsequent related papers such as Lisa Randall and Raman Sundrum, Phys. Rev. Lett., **83,** 3370 (1999).

[8] H. Weyl, Z. Physik. **56**, 330 (1929).

[9] R. Utiyama, Phys. Rev. **101**, 1597 (1956), T. W. B. Kibble, Jour. Math. Phys. **2**, 212 (1961), J. Schwinger, Phys. Rev. **130**, 800 (1963), J. Schwinger, Phys. Rev. **130**, 1253 (1963). **We generally use their notation in this chapter.**

[10] C. Isham, A. Salam, and J. Strathdee, Phys. Rev. D3, 867 (1971); ________,Lett. Al Nuov. Cim., **5**, 969 (1972) and references therein.

[11] Typographic note: Since Times Roman type does not distinguish between 1 (one) and l (el) the vierbein l is distinguished by indices or appearance in a matrix equation.

3. It is convenient to establish a matrix form of a vierbein with

$$l^{\mu}(x) = l^{\mu a}(x)\gamma_a \qquad (4.2)$$

where the four matrices γ_a are the familiar Dirac matrices.

4. Under a local Special Relativistic transformation S a spin ½ field transforms as

$$\psi(x) \rightarrow S(x)\psi(x) \qquad (4.3)$$

and the matrix form of a vierbein transforms as

$$l^{\mu}(x) \rightarrow S(x)l^{\mu}(x)S^{-1}(x) \qquad (4.4)$$

5. Under a General Relativistic transformation a vierbein transforms as

$$l'^{\mu a}(x') = \partial x_{\nu}/\partial x'_{\mu}\, l^{\nu a}(x) \qquad (4.5)$$

Other features of the vierbein formalism can be found in papers in the aforementioned footnotes and in Weinberg (1972) as well as other books.
We will assume that the x coordinate system is complex-valued initially. Then we will map it to real values in keeping with Blaha (2015a) and our other books. We showed many years ago that faster-than-light motion is possible and that it requires complex-valued coordinates and complex Lorentz transformations.

4.2 Extension of the Vierbein Formalism to Incorporate the Strong Interaction

We can simply extend[12] the above General Relativistic vierbein formalism to include the the SU(3) Strong Interaction by adding an additional SU(3) index to the vierbein.[13] Thus the extended formalism includes

1. An SU(3) extended vierbein (tetrad) has the form $l^{\mu a i}(x)$ where the additional index i for SU(3) ranges from 1 through 8.

2. The curve space-time metric tensor now is defined by

[12] This extension, with a slight modification, appears in the author's Gravity Research Foundation 1976 Essay competition submission for which he received Honorable Mention.
[13] We do not extend the *vierbein* formalism to include the Weak interaction because it appears to have a fundamentally different character because of its intimate connection with Parity violation.

$$g_{\mu\nu} = \eta_{ab} l_\mu{}^{ai}(x) l_\nu{}^b{}_i(x) \tag{4.6}$$

with a sum over i.

3. It is again convenient to establish a matrix form of a vierbein with

$$l^\mu(x) = l^{\mu ai}(x)\gamma_a T_i \tag{4.7}$$

where the eight matrices T_i are SU(3) generators in the fundamental $\underline{3}$ representation.
4. Under an SU(3) local gauge transformation C the *unified* vierbein, in matrix form, transforms as

$$l^\mu(x) \rightarrow C(x) l^\mu(x) C^{-1}(x) \tag{4.8}$$

Transformations under General Relativity and Special Relativity have the same form as above.

4.3 Physical Interpretation of the SU(3) Extended Vierbein

The extended vierbein $l^{\mu ai}(x)$ can be viewed as located at a point in a 32-dimensional complex-valued space.

$$l^{\mu ai}(x) = (\partial \xi_X{}^{ai}(x)/\partial x_\mu)_{X=h(x)} \tag{4.9}$$

where $\xi_X{}^{ai}$ is a set of locally inertial coordinates located at a 32-dimensional point X, and x = h(x) is a 4-dimensional point in a tangent subspace of the 32-dimensional space:

$$X = h(x)$$

The relation between complex 4-dimensional coordinates x and the 32-dimensional coordinates X is an embedding of a 4-dimensional surface within a 32-dimensional complex space when account is taken of the range of possible x values. We have considered such embeddings in Blaha (2015a), and in earlier books, and developed a theory of a 16-dimensional complex-valued space (the *Megaverse*) that contains our universe and probably many other universes. The study, in which we are now engaged, adds to the reasons given in earlier books for belief in the Megaverse.[14]

---

[14] Many years ago, the term Megaverse was introduced by William James. It signifies a space with many resident universes. The word multiverse was popularized by the work of Everritt, and others, and signifies many parallel universes based on quantum mechanical considerations.

4.4 The Megaverse – The 16-dimensional Space with Complex-valued Coordinates

The 32-dimensional flat, complex-valued, tangent space considered above can be mapped onto a flat 16-dimensional universe with complex-valued coordinates in a straightforward way. Upon making that transition we can turn the above process around and view our curved 4-dimensional universe as a flat complex-valued subspace residing within a 16-dimensional, complex-valued, tangent space now grown to a full space by extension to all possible points X.[15] This is the *Megaverse*.

Since the coordinates of our space-time are necessarily complex-valued (Blaha (2015a) and earlier books), it is natural for the Megaverse to have complex coordinates as well.

The Megaverse was considered flat in the previous section. Now we can introduce interactions (and a definition of mass-energy) that make the Megaverse curved as well.[16]

Thus we view the Megaverse as a curved, complex-valued, 16-dimensional space containing universes interacting with each other through long-range forces such as the elusive Baryon Number force and the Dark Baryon Number force.[17] Universes then appear to be strewn through the Megaverse like galaxies within our universe. Universes may collide and perhaps combine over incredibly long time frames. Blaha (2015a) and (2015b) considers universe dynamics in some detail and proposes a quantum field theory of universe particles, *uons*, based on the form of the Wheeler-DeWitt equation. The reader is referred to these works for further discussion. We return to the consideration of the unification of Gravitation and the Strong Interaction.

A complex-valued, 4-dimensional curved universe can be mapped to a surface in a 10-dimensional complex 'Euclidean' space since the metric tensor has 10 components. For a universe with 10 complex-valued metric tensor components a complex-valued Euclidean space must have at least 20 dimensions or ten complex-valued dimensions. Since the Megaverse has 16 complex-valued dimensions our universe, as well as other 4-dimensional universes, can be mapped to it as surfaces using maps such as $\mathbf{X} = \mathbf{h}(\mathbf{x})$ where $\mathbf{X}$ is a 16-vector, $\mathbf{x}$ is a 4-vector and $\mathbf{h}()$ is a a 16-vector function.

4.5 Dynamics of Unification

This section describes the basic fields and their dynamics.

---

[15] An informed reader may ask: If the Strong Interactions are confined as they appear to be, how can you define a space which, in part, has an SU(3) Strong Interaction part? Would not the space be 'confined' in some way? The answer lies in realizing that coordinates are not confined. Particles are confined. Coordinates are not fields, and the metric and the curvature tensor of the space are independent of the Strong Interaction outside of regions containing quarks and gluons. Within a hadron, or quark-gluon plasma, the interplay of Gravitation and the Strong Interaction will have effects on the curvature of space.

[16] See Blaha (2015a) and (2015b).

[17] We discuss these interactions, and other interactions, in chapter 10.

4.5.1 The ElectroMagnetic Field

The U(1) electromagnetic gauge fields[18] are defined as $A_E^{1\mu}(x)$ and $A_E^{2\mu}(x)$.[19] Under a local electromagnetic gauge transformation $C_E(x)$ the gauge fields transform as

$$A_E^{1\mu}(x) \rightarrow C_E(x)A_E^{1\mu}(x)C_E^{-1}(x) - i\, C_E(x)\partial^\mu C_E^{-1}(x) \tag{4.10a}$$

and

$$A_E^{2\mu}(x) \rightarrow C_E(x)A_E^{2\mu}(x)C_E^{-1}(x) \tag{4.10b}$$

4.5.2 The Gravitation, Weak, and Strong Interaction Fields

The spinor connection used in formulations of vierbein gravity is $B^1{}_{\mu ab}(x)$ where a and b are tangent space indices. The vector is combined with γ matrices for use in matrix equations:

$$B^{1\mu} = B^{1\mu}{}_{ab}\Sigma^{ab} \tag{4.11}$$

where

$$\Sigma^{ab} = i\,[\gamma^a, \gamma^b]/4 \tag{4.12}$$

Under a local Lorentz transformation S

$$B^{1\mu}(x) \rightarrow S(x)B^{1\mu}(x)S^{-1}(x) - i\, S(x)\partial^\mu S^{-1}(x) \tag{4.13}$$

Similarly a spin ½ field transforms as

$$(\partial^\mu + i\, B^{1\mu})\psi \rightarrow S(\partial^\mu + i\, B^{1\mu})\psi \tag{4.14}$$

The SU(3) gauge fields are defined as $A^{1i\mu}(x)$ and $A^{2i\mu}(x)$ for i = 1, … , 8. Using SU(3) generators we define the matrix form by $A^\mu(x) = A^{i\mu}(x)T_i$. Under an SU(3) gauge transformation C the gauge field transforms as

$$A^{1\mu}(x) \rightarrow C(x)A^{1\mu}(x)C^{-1}(x) - i\, C(x)\partial^\mu C^{-1}(x) \tag{4.15a}$$

and

$$A^{2\mu}(x) \rightarrow C(x)A^{2\mu}(x)C^{-1}(x) \tag{4.15b}$$

The Weak interaction SU(2) gauge fields are defined as $W^{1i\mu}(x)$ and $W^{2i\mu}(x)$ for i = 1, 2, 3. Using SU(2) generators we define the matrix form by $W^{k\mu}(x) = W^{ki\mu}(x)\tau_i$ for k= 1, 2. Under an SU(2) gauge transformation C_W the gauge field transforms as

$$W^{1\mu}(x) \rightarrow C_W(x)W^{1\mu}(x)C_W^{-1}(x) - i\, C_W(x)\partial^\mu C_W^{-1}(x) \tag{4.16a}$$

---

[18] We introduce two fields as we did in our article S. Blaha, Phys. Rev. D**10**, 4268 (July, 1974). These fields enable us to define a free electromagnetic lagrangian that is linear in the fields for reasons given elsewhere.

[19] See subsection 4.5.4 for descriptions of the secondary $A_E^{2\mu}(x)$, $A^{2i\mu}(x)$, and $W^{2i\mu}(x)$ gauge fields. Subsection 4.5.3 describes the secondary $B'^\sigma{}_{\nu\mu}$ affine connecton.

and

$$W^{2\mu}(x) \rightarrow C_W(x)W^{2\mu}(x)C_W^{-1}(x) \tag{4.16b}$$

4.5.3 The Gravitation Affine Connection

The affine connection is most often viewed as a derived quantity—part of the derivation of the curvature tensor in General Relativity. It is typically derived from manipulations of the metric $g_{\mu\nu}$. However, the affine connection can also be viewed as a set of independent fields that become related to the metric via dynamic equations.

Some years ago A. Einstein and H. Weyl[20] pointed out that the metric and the affine connection should be treated as independent quantities and subject to independent arbitrary infinitesimal variations:

> "In contrast to Einstein's original "metric" conception in terms of the $g_{\nu\mu}$ there was later developed, by Eddington, by Einstein himself, and recently by Schrödinger, an affine field theory operating with the components $\Gamma^\sigma{}_{\nu\mu}$ of an affine connection. But in 1925 Einstein also advocated a "mixed" formulation by means of a lagrangian in which both the $g_{\nu\mu}$ and the $\Gamma^\sigma{}_{\nu\mu}$ are taken as basic field quantities and submitted to independent arbitrary infinitesimal variations.[21] In certain respects this seems to be the most natural procedure."

Following this approach we introduce affine connections. For the gravitational part of the vierbein we take the spinor affine connection to be given in eq. 4.11.

Further we define another affine connection, which is related to the usual General Relativistic gravitational connection $\Gamma^\sigma{}_{\nu\mu}$:

$$B'^{\sigma}{}_{\nu\mu} = B'^{\sigma ab}{}_{\nu\mu}\Sigma_{ab} \tag{4.17}$$

with Σ_{ab} defined by eq. 4.12. $B'^{\sigma}{}_{\nu\mu}$ will be determined by the dynamic equations.

Under a local Lorentz transformation S(x) we require both $B'^{\sigma}{}_{\nu\mu}$ and $B'^{\sigma ab}{}_{\nu\mu}$ transform homogeneously

$$\begin{aligned} B'^{\sigma ab}{}_{\nu\mu} &\rightarrow S(x)B'^{\sigma ab}{}_{\nu\mu}S^{-1}(x) \\ B'^{\sigma}{}_{\nu\mu} &\rightarrow S(x)B'^{\sigma}{}_{\nu\mu}S^{-1}(x) \end{aligned} \tag{4.18}$$

We futher define $B'^{\sigma}{}_{\nu\mu}$ as

$$B'^{\alpha ab}{}_{\nu\mu}\, l_{\alpha a}{}^{i}(x) l^{\sigma}{}_{bi}(x) = i\,\Gamma^\sigma{}_{\nu\mu}/2 \tag{4.19}$$

with an implicit sum over i.

Thus

$$B'^{\sigma}{}_{\nu\mu}\times l_\sigma \equiv [B'^{\sigma}{}_{\nu\mu}, l_\sigma] = \Gamma^\sigma{}_{\nu\mu} l_\sigma \tag{4.20}$$

---

[20] H. Weyl, Phys. Rev. **77**, 699 (1950).
[21] A. Einstein, Sitzungsber., Preuss. Akad. Der Wissensch. (1925), p. 414.

for all implicit SU(3) indices i.

4.5.4 The SU(3), SU(2), and Electromagnetic Affine Connections

We now define an additional SU(3) Yang-Mills affine connection (a general coordinate transformation tensor) as:

$$A'^{\sigma}{}_{\nu\mu} = A'^{\sigma i}{}_{\nu\mu}T_i \tag{4.21}$$

Under a local gauge transformation C(x) we define its gauge transformation to be homogeneous:

$$A'^{\sigma}{}_{\nu\mu}(x) \rightarrow C(x)A'^{\sigma}{}_{\nu\mu}(x)C^{-1}(x) \tag{4.22}$$

Based on the above stated views of Einstein and Weyl that the metric and the affine connection should be treated as independent quantities and subject to independent arbitrary infinitesimal variations we define the SU(3) affine connection in terms of a new SU(3) gauge field A'_μ as

$$A'^{\sigma}{}_{\nu\mu} = g^{\sigma}{}_{\nu}A^{2}{}_{\mu} + g^{\sigma}{}_{\mu}A^{2}{}_{\nu} \tag{4.23}$$

where

$$A^{2}{}_{\mu} = A^{2i}{}_{\mu}T_i \tag{4.24}$$

Eq. 4.23 maintains the usual affine connection symmetry in ν and μ of $A'^{\sigma}{}_{\nu\mu}$. Eq. 4.22 implies that A'_μ is a Lorentz vector that transforms homogeneously under local SU(3) gauge transformations C(x).

$$A^{2\mu}(x) \rightarrow C(x)A^{2\mu}(x)C^{-1}(x) \tag{4.25}$$

Similarly we define an additional electromagnetic affine connection $A_E{}'^{\sigma}{}_{\nu\mu}(x)$ which is a general coordinate transformation tensor. Under a local electromagnetic gauge transformation $C_E(x)$ we define its gauge transformation to be homogeneous:

$$A_E{}'^{\sigma}{}_{\nu\mu}(x) \rightarrow C_E(x)A_E{}'^{\sigma}{}_{\nu\mu}(x)C_E{}^{-1}(x) \tag{4.26}$$

Again we treat it as an independent quantity and subject to independent arbitrary infinitesimal variations. We define this affine connection in terms of a new gauge field $A_E{}'_\mu$ as

$$A_E{}'^{\sigma}{}_{\nu\mu} = g^{\sigma}{}_{\nu}A_E{}^{2}{}_{\mu} + g^{\sigma}{}_{\mu}A_E{}^{2}{}_{\nu} \tag{4.27}$$

where

$$A_E{}^{2}{}_{\mu} = A_E{}^{2i}{}_{\mu}T_i \tag{4.28}$$

Eq. 4.27 maintains the usual affine connection symmetry in ν and μ of $A'^{\sigma}{}_{\nu\mu}$. Eq. 4.26 implies that $A_E{}'_\mu$ is a Lorentz vector that transforms homogeneously under local electromagnetic gauge transformations $C_E(x)$.

$$A_E^{2\mu}(x) \to C_E(x)A_E^{2\mu}(x)C_E^{-1}(x) \qquad (4.29)$$

We now define a Weak SU(2) Yang-Mills affine connection (a general coordinate transformation tensor) as:

$$W'^{\sigma}{}_{\nu\mu} = W'^{\sigma i}{}_{\nu\mu}\tau_i \qquad (4.30)$$

Under a local gauge transformation $C_W(x)$ we define its gauge transformation to be homogeneous:

$$W'^{\sigma}{}_{\nu\mu}(x) \to C_W(x)W'^{\sigma}{}_{\nu\mu}(x)C_W^{-1}(x) \qquad (4.31)$$

Based on the above stated views of Einstein and Weyl that the metric and the affine connection should be treated as independent quantities and subject to independent arbitrary infinitesimal variations we define the Weak SU(2) affine connection in terms of the new Weak SU(2) gauge field W'_μ as

$$W'^{\sigma}{}_{\nu\mu} = g^{\sigma}{}_{\nu}W^2{}_{\mu} + g^{\sigma}{}_{\mu}W^2{}_{\nu} \qquad (4.32)$$

where

$$W^2{}_{\mu} = W^{2i}{}_{\mu}\tau_i \qquad (4.33)$$

Eq. 4.32 maintains the usual affine connection symmetry in ν and μ of $W'^{\sigma}{}_{\nu\mu}$. W'_μ is a Lorentz vector that transforms homogeneously under local SU(2) gauge transformations $C_W(x)$.

$$W^{2\mu}(x) \to C_W(x)W^{2\mu}(x)C_W^{-1}(x) \qquad (4.34)$$

4.5.5 Covariant Derivatives and the Unified Curvature Tensor

In this subsection we define the covariant derivative, curvature tensor, Ricci tensor, and curvature scalar for the (ElectroGraviWeakStrong) unification of electromagnetism, gravity, the Weak interactions, and Strong SU(3) in a 32-dimensional complex space without the use of the subsidiary affine connections defined by eqs. 4.17 and 4.21.

Based on the preceding sections we use the following covariant derivative of a vector in 32-dimensional space:[22, 23]

$$\begin{aligned} D_\nu V_\mu &= (\partial_\nu + iB^1{}_\nu + iA^1{}_\nu + iA_E{}^1{}_\nu + iW^1{}_\nu)V_\mu - \Gamma^{\sigma}{}_{\nu\mu}V_\sigma \qquad (4.35) \\ &= (\partial_\nu + iC_\nu)V_\mu - \Gamma^{\sigma}{}_{\nu\mu}V_\sigma \\ &= [g^{\sigma}{}_{\mu}\partial_\nu + ig^{\sigma}{}_{\mu}C_\nu - \Gamma^{\sigma}{}_{\nu\mu}]V_\sigma \end{aligned}$$

---

[22] We use the superscript '1' to distinguish from secondary connections introduced in the following sections,

[23] It is understood that $\partial_\nu + iA_E{}^1{}_\nu$ have implicit spinor, SU(2), and SU(3) identity matrix factors, $B^1{}_\nu$ has implicit SU(3) and SU(2) identity matrix factors, $W^1{}_\nu$ has implicit SU(3) and spinor identity matrix factors, and $A^1{}_\nu$ has implicit spinor and SU(2) identity matrix factors both here and in the following equations.

$$= [g^{\sigma}{}_{\mu}\partial_{\nu} + iD^{\sigma}{}_{\mu\nu}]V_{\sigma}$$

where

$$C_{\mu} = B^{1}{}_{\mu} + A_{E}{}^{1}{}_{\mu} + A^{1}{}_{\mu} + W^{1}{}_{\mu} \quad (4.36)$$
$$D^{\sigma}{}_{\mu\nu} = g^{\sigma}{}_{\mu}C_{\nu} + i\Gamma^{\sigma}{}_{\nu\mu} \quad (4.37)$$

Eq. 4.35 enables us to simply derive the Riemann-Christoffel curvature tensor, and then its contractions $R_{\mu\nu}$ and R using

$$(D_{\nu}D_{\mu} - D_{\mu}D_{\nu})V_{\sigma} = R^{\beta}{}_{\sigma\nu\mu}V_{\beta} \quad (4.38)$$

where

$$V_{\sigma} = V_{\sigma}{}^{ai}(x)\gamma_{a}T_{i} \quad (4.39)$$

Then

$$D_{\nu}D_{\mu}V_{\sigma} = \{g^{\alpha}{}_{\mu}(\partial_{\nu} + iB^{1}{}_{\nu} + iA_{E}{}^{1}{}_{\nu} + iA^{1}{}_{\nu} + iW^{1}{}_{\nu}) - \Gamma^{\alpha}{}_{\mu\nu}\}\{g^{\beta}{}_{\sigma}(\partial_{\alpha} + iB^{1}{}_{\alpha} + iA_{E}{}^{1}{}_{\alpha} + iA^{1}{}_{\alpha} + iW^{1}{}_{\alpha})V_{\beta} - \Gamma^{\beta}{}_{\sigma\alpha}V_{\beta}\} -$$
$$- \Gamma^{\gamma}{}_{\nu\sigma}\{g^{\alpha}{}_{\gamma}(\partial_{\mu} + iB^{1}{}_{\mu} + iA_{E}{}^{1}{}_{\mu} + iA^{1}{}_{\mu} + iW^{1}{}_{\mu})V_{\alpha} - \Gamma^{\alpha}{}_{\gamma\mu}V_{\alpha}\} \quad (4.40)$$

and

$$R^{\beta}{}_{\sigma\nu\mu}V_{\beta} = g^{\alpha}{}_{\mu}(\partial_{\nu} + iB^{1}{}_{\nu} + iA_{E}{}^{1}{}_{\nu} + iA^{1}{}_{\nu} + iW^{1}{}_{\nu})g^{\beta}{}_{\sigma}(\partial_{\alpha} + iB^{1}{}_{\alpha} + iA_{E}{}^{1}{}_{\alpha} + iA^{1}{}_{\alpha} + iW^{1}{}_{\alpha})V_{\beta} -$$
$$- \Gamma^{\alpha}{}_{\mu\nu}g^{\beta}{}_{\sigma}(\partial_{\alpha} + iB^{1}{}_{\alpha} + iA_{E}{}^{1}{}_{\alpha} + iA^{1}{}_{\alpha} + iW^{1}{}_{\alpha})V_{\beta} +$$
$$+ \Gamma^{\alpha}{}_{\mu\nu}\Gamma^{\beta}{}_{\sigma\alpha}V_{\beta} - g^{\alpha}{}_{\mu}(\partial_{\nu} + iB^{1}{}_{\nu} + iA_{E}{}^{1}{}_{\nu} + iA^{1}{}_{\nu} + iW^{1}{}_{\nu})\Gamma^{\beta}{}_{\sigma\alpha}V_{\beta} -$$
$$- \Gamma^{\gamma}{}_{\nu\sigma}\{g^{\alpha}{}_{\gamma}(\partial_{\mu} + iB^{1}{}_{\mu} + iA_{E}{}^{1}{}_{\mu} + iA^{1}{}_{\mu} + iW^{1}{}_{\mu})V_{\alpha} - \Gamma^{\alpha}{}_{\gamma\mu}V_{\alpha}\} -$$
$$- \{\mu \leftrightarrow \nu\}$$

$$= ig^{\beta}{}_{\sigma}F^{1}{}_{\nu\mu}V_{\beta} + (ig^{\beta}{}_{\sigma}B^{1}{}_{\nu\mu} + \partial_{\mu}\Gamma^{\beta}{}_{\sigma\nu} - \partial_{\nu}\Gamma^{\beta}{}_{\sigma\mu} + \Gamma^{\gamma}{}_{\nu\sigma}\Gamma^{\beta}{}_{\gamma\mu} - \Gamma^{\gamma}{}_{\mu\sigma}\Gamma^{\beta}{}_{\gamma\nu})V_{\beta}$$
$$= R_{E}{}^{\beta}{}_{\sigma\nu\mu}V_{\beta} + R_{SU(3)}{}^{\beta}{}_{\sigma\nu\mu}V_{\beta} + R_{SU(2)}{}^{\beta}{}_{\sigma\nu\mu}V_{\beta} + R^{\beta}{}_{\sigma\nu\mu}V_{\beta} \quad (4.41)$$

where

$$F^{1}{}_{\kappa\mu} = \partial A_{E}{}^{1}{}_{\mu}/\partial x^{\kappa} - \partial A_{E}{}^{1}{}_{\kappa}/\partial x^{\mu} + \partial A^{1}{}_{\mu}/\partial x^{\kappa} - \partial A^{1}{}_{\kappa}/\partial x^{\mu} + i[A^{1}{}_{\kappa}\times A^{1}{}_{\mu}] + \partial W^{1}{}_{\mu}/\partial x^{\kappa} - \partial W^{1}{}_{\kappa}/\partial x^{\mu} + i[W^{1}{}_{\kappa}\times W^{1}{}_{\mu}]$$
$$= F_{E}{}^{1}{}_{\kappa\mu} + F^{1}{}_{\kappa\mu} + F_{W}{}^{1}{}_{\kappa\mu} \quad (4.42)$$

with

$$F_{E}{}^{1}{}_{\kappa\mu} = \partial A_{E}{}^{1}{}_{\mu}/\partial x^{\kappa} - \partial A_{E}{}^{1}{}_{\kappa}/\partial x^{\mu} \quad (4.42a)$$
$$F^{1}{}_{\kappa\mu} = \partial A^{1}{}_{\mu}/\partial x^{\kappa} - \partial A^{1}{}_{\kappa}/\partial x^{\mu} + i[A^{1}{}_{\kappa}\times A^{1}{}_{\mu}] \quad (4.42b)$$
$$F_{W}{}^{1}{}_{\kappa\mu} = \partial W^{1}{}_{\mu}/\partial x^{\kappa} - \partial W^{1}{}_{\kappa}/\partial x^{\mu} + i[W^{1}{}_{\kappa}\times W^{1}{}_{\mu}] \quad (4.42c)$$

and

$$B^{1}{}_{\kappa\mu} = \partial B^{1}{}_{\mu}/\partial x^{\kappa} - \partial B^{1}{}_{\kappa}/\partial x^{\mu} + iB^{1}{}_{\kappa}\times B^{1}{}_{\mu} - iB^{1}{}_{\mu}\times B^{1}{}_{\kappa} \quad (4.42d)$$

Thus the total electromagnetic, SU(3) and Gravity curvature tensor is

$$R_{tot}{}^{\beta}{}_{\sigma\nu\mu} = ig^{\beta}{}_{\sigma}F_E{}^1{}_{\nu\mu} + ig^{\beta}{}_{\sigma}F^1{}_{\nu\mu} + ig^{\beta}{}_{\sigma}F_W{}^1{}_{\nu\mu} + (ig^{\beta}{}_{\sigma}B^1{}_{\nu\mu} + \partial_\mu\Gamma^{\beta}{}_{\sigma\nu} - \partial_\nu\Gamma^{\beta}{}_{\sigma\mu} + \Gamma^{\gamma}{}_{\nu\sigma}\Gamma^{\beta}{}_{\gamma\mu} - \Gamma^{\gamma}{}_{\mu\sigma}\Gamma^{\beta}{}_{\gamma\nu})$$
$$= R_E{}^{\beta}{}_{\sigma\nu\mu} + R_{SU(3)}{}^{\beta}{}_{\sigma\nu\mu} + R_{SU(2)}{}^{\beta}{}_{\sigma\nu\mu} + R^{\beta}{}_{\sigma\nu\mu} \qquad (4.43)$$

Note $R_{tot}{}^{\beta}{}_{\sigma\nu\mu}$ factorizes into an electromagnetic part, an SU(3) part, and a Riemann-Christoffel curvature tensor part:

$$R_E{}^{\beta}{}_{\sigma\nu\mu} = ig^{\beta}{}_{\sigma}F_E{}^1{}_{\nu\mu} \qquad (4.44a)$$

$$R_{SU(3)}{}^{\beta}{}_{\sigma\nu\mu} = ig^{\beta}{}_{\sigma}F^1{}_{\nu\mu} \qquad (4.44b)$$

$$R_{SU(2)}{}^{\beta}{}_{\sigma\nu\mu} = ig^{\beta}{}_{\sigma}F_W{}^1{}_{\nu\mu} \qquad (4.44c)$$

$$R^{\beta}{}_{\sigma\nu\mu} = ig^{\beta}{}_{\sigma}B^1{}_{\nu\mu} + \partial_\mu\Gamma^{\beta}{}_{\sigma\nu} - \partial_\nu\Gamma^{\beta}{}_{\sigma\mu} + \Gamma^{\gamma}{}_{\nu\sigma}\Gamma^{\beta}{}_{\gamma\mu} - \Gamma^{\gamma}{}_{\mu\sigma}\Gamma^{\beta}{}_{\gamma\nu} \qquad (4.45)$$

$R_{tot}{}^{\beta}{}_{\sigma\nu\mu}$ is the Riemann-Christoffel curvature tensor for the complex 32-dimensional space generated by SU(3) and 4-dimensional General Coordinate transformations. $R^{\beta}{}_{\sigma\nu\mu}$ is the Riemann-Christoffel curvature tensor for our complex 4-dimensional space-time. Note the anti-symmetry in μ and ν.

The total Ricci tensor is

$$R_{tot\sigma\mu} = R_{tot}{}^{\beta}{}_{\sigma\beta\mu}$$
$$= iF_E{}^1{}_{\sigma\mu} + iF^1{}_{\sigma\mu} + iF_W{}^1{}_{\sigma\mu} + (iB^1{}_{\sigma\mu} + \partial_\mu\Gamma^{\beta}{}_{\sigma\beta} - \partial_\beta\Gamma^{\beta}{}_{\sigma\mu} + \Gamma^{\gamma}{}_{\beta\sigma}\Gamma^{\beta}{}_{\gamma\mu} - \Gamma^{\gamma}{}_{\mu\sigma}\Gamma^{\beta}{}_{\gamma\beta})$$
$$= R_{E\sigma\mu} + R_{SU(3)\sigma\mu} + R_{SU(2)\sigma\mu} + R_{\sigma\mu} \qquad (4.46)$$

The curvature scalar is

$$R_{tot} = g^{\sigma\mu}R_{tot\sigma\mu} = g^{\sigma\mu}(\partial_\mu\Gamma^{\beta}{}_{\sigma\beta} - \partial_\beta\Gamma^{\beta}{}_{\sigma\mu} + \Gamma^{\gamma}{}_{\beta\sigma}\Gamma^{\beta}{}_{\gamma\mu} - \Gamma^{\gamma}{}_{\mu\sigma}\Gamma^{\beta}{}_{\gamma\beta}) \qquad (4.47)$$

Eq. 4.46 is the Ricci tensor for the complex 32-dimensional space generated by SU(3) and 4-dimensional complex General Coordinate transformations. Eq. 4.47 is its curvature scalar. Note the curvature scalar is independent of electromagnetism, SU(3), SU(2), and the spinor connection $B^1{}_\mu$.

4.5.6 Covariant Derivatives and the Unified Curvature Tensor with the Additional Electromagnetic, SU(3) and Vierbein Affine Connections

In this subsection we define the covariant derivative, curvature tensor, Ricci tensor, and curvature scalar for the unification of electromagnetism, gravity, SU(2), and SU(3) in a 32-dimensional complex space with the use of the subsidiary electromagnetic, Weak SU(2), and Strong SU(3) affine connections defined above, together with a secondary metric, spinor connection, and gravitational affine connection. In accord with Einstein and Weyl we will subject these secondary quantities to independent variations when deriving the field equations.

The primary motivation for these secondary fields is to enable us to use the canonical lagrangian formalism for higher derivative, 4th order SU(3) and gravitation dynamic terms. *These secondary quantities will enable us to define a canonical lagrangian formulation of gravity unified with electromagnetism, Weak SU(2), and Strong SU(3) that yields a modified form of the gravitational potential, and an SU(3) interaction with a color confining r potential.*

We begin by defining a subsidiary gravitational metric to support a higher derivative, lagrangian formulation:

$$g^{2\mu\nu} = g^{2\nu\mu} \tag{4.48}$$
$$B^{2\mu} = B^{2\mu}{}_{ab}\Sigma^{ab} \tag{4.49}$$

where Σ^{ab} is defined by eq. 4.12, and a secondary affine connection:

$$\Gamma^{2\lambda}{}_{\mu\nu} = \tfrac{1}{2} g^{2\lambda\alpha}(\partial_\mu g^2{}_{\alpha\nu} + \partial_\nu g^2{}_{\alpha\mu} - \partial_\alpha g^2{}_{\mu\nu}) \tag{4.50}$$

We use the following generalized covariant derivative of a vector in our 32-dimensional space:[24]

$$\begin{aligned} D_\nu V_\mu &= (\partial_\nu + iF_\nu)V_\mu - H^\sigma{}_{\nu\mu}V_\sigma \\ &= [g^\sigma{}_\mu\partial_\nu + ig^\sigma{}_\mu F_\nu - H^\sigma{}_{\nu\mu}]V_\sigma \\ &= [g^\sigma{}_\mu\partial_\nu + iD^\sigma{}_{\mu\nu}]V_\sigma \end{aligned} \tag{4.51}$$

where

$$F_\mu = B^1{}_\mu + A_E{}^1{}_\mu + W^1{}_\mu + A^1{}_\mu + B^2{}_\mu + A_E{}^2{}_\mu + W^2{}_\mu + A^2{}_\mu \tag{4.52}$$
$$H^\sigma{}_{\nu\mu} = \Gamma^\sigma{}_{\nu\mu} + \Gamma^{2\sigma}{}_{\nu\mu} \tag{4.53}$$
$$D^\sigma{}_{\mu\nu} = g^\sigma{}_\mu F_\nu + iH^\sigma{}_{\nu\mu} \tag{4.54}$$

Eq. 4.51 enables us to derive the Riemann-Christoffel curvature tensor, and then its contractions $R_{\mu\nu}$ and R using

$$(D_\nu D_\mu - D_\mu D_\nu)V_\sigma = R^\beta{}_{\sigma\nu\mu}V_\beta \tag{4.55}$$

where

$$V_\sigma = V_\sigma{}^{ai}(x)\gamma_a T_i \tag{4.56}$$

Then

$$D_\nu D_\mu V_\sigma = \{g^\alpha{}_\mu(\partial_\nu + iF_\nu) - H^\alpha{}_{\mu\nu}\}\{g^\beta{}_\sigma(\partial_\alpha + iF_\alpha)V_\beta - H^\beta{}_{\sigma\alpha}V_\beta\} - H^\gamma{}_{\nu\sigma}\{g^\alpha{}_\gamma(\partial_\mu + iF_\mu)V_\alpha - H^\alpha{}_{\gamma\mu}V_\alpha\} \tag{4.57}$$

As a result

$$\begin{aligned} R'^\beta{}_{\sigma\nu\mu}V_\beta = {} & g^\alpha{}_\mu(\partial_\nu + iF_\nu)g^\beta{}_\sigma(\partial_\alpha + iF_\alpha)V_\beta - H^\alpha{}_{\mu\nu}g^\beta{}_\sigma(\partial_\alpha + iF_\alpha)V_\beta + \\ & + H^\alpha{}_{\mu\nu}H^\beta{}_{\sigma\alpha}V_\beta - g^\alpha{}_\mu(\partial_\nu + iF_\nu)H^\beta{}_{\sigma\alpha}V_\beta - H^\gamma{}_{\nu\sigma}\{g^\alpha{}_\gamma(\partial_\mu + iF_\mu)V_\alpha - H^\alpha{}_{\gamma\mu}V_\alpha\} - \\ & - \{\mu\leftrightarrow\nu\} \end{aligned}$$

[24] We use the superscript '1' to distinguish primary connections from secondary connections labeled '2'.

$$= ig^{\beta}{}_{\sigma}(\partial_\nu F_\mu - \partial_\mu F_\nu - i[F_\nu, F_\mu])V_\beta + (\partial_\mu H^{\beta}{}_{\sigma\nu} - \partial_\nu H^{\beta}{}_{\sigma\mu} + H^{\gamma}{}_{\nu\sigma}H^{\beta}{}_{\gamma\mu} - H^{\gamma}{}_{\mu\sigma}H^{\beta}{}_{\gamma\nu})V_\beta$$
$$= ig^{\beta}{}_{\sigma}(F_E{}^1{}_{\nu\mu} + F_E{}^2{}_{\nu\mu} + F_W{}^1{}_{\nu\mu} + F_W{}^2{}_{\nu\mu} + F^1{}_{\nu\mu} + F^2{}_{\nu\mu})V_\beta +$$
$$+ (ig^{\beta}{}_{\sigma}B^1{}_{\nu\mu} + ig^{\beta}{}_{\sigma}B^2{}_{\nu\mu} + \partial_\mu H^{\beta}{}_{\sigma\nu} - \partial_\nu H^{\beta}{}_{\sigma\mu} + H^{\gamma}{}_{\nu\sigma}H^{\beta}{}_{\gamma\mu} - H^{\gamma}{}_{\mu\sigma}H^{\beta}{}_{\gamma\nu})V_\beta$$

$$= R'_E{}^{\beta}{}_{\sigma\nu\mu}V_\beta + R'_{SU(2)}{}^{\beta}{}_{\sigma\nu\mu}V_\beta + R'_{SU(3)}{}^{\beta}{}_{\sigma\nu\mu}V_\beta + R'_G{}^{\beta}{}_{\sigma\nu\mu}V_\beta \qquad (4.58)$$

where

$$R'_{SU(3)}{}^{\beta}{}_{\sigma\nu\mu} = ig^{\beta}{}_{\sigma}(F^1{}_{\nu\mu} + F^2{}_{\nu\mu}) \qquad (4.59)$$
$$R'_{SU(2)}{}^{\beta}{}_{\sigma\nu\mu} = ig^{\beta}{}_{\sigma}(F_W{}^1{}_{\nu\mu} + F_W{}^2{}_{\nu\mu})$$
$$R'_E{}^{\beta}{}_{\sigma\nu\mu} = ig^{\beta}{}_{\sigma}(F_E{}^1{}_{\nu\mu} + F_E{}^2{}_{\nu\mu})$$

and

$$R'_G{}^{\beta}{}_{\sigma\nu\mu} = ig^{\beta}{}_{\sigma}(B^1{}_{\nu\mu} + B^2{}_{\nu\mu}) + \partial_\mu\Gamma^{\beta}{}_{\sigma\nu} - \partial_\nu\Gamma^{\beta}{}_{\sigma\mu} + \Gamma^{\gamma}{}_{\nu\sigma}\Gamma^{\beta}{}_{\gamma\mu} - \Gamma^{\gamma}{}_{\mu\sigma}\Gamma^{\beta}{}_{\gamma\nu} + \partial_\mu\Gamma^{2\beta}{}_{\sigma\nu} - \partial_\nu\Gamma^{2\beta}{}_{\sigma\mu} +$$
$$+\Gamma^{2\gamma}{}_{\nu\sigma}\Gamma^{2\beta}{}_{\gamma\mu} - \Gamma^{2\gamma}{}_{\mu\sigma}\Gamma^{2\beta}{}_{\gamma\nu} + \Gamma^{\gamma}{}_{\nu\sigma}\Gamma^{2\beta}{}_{\gamma\mu} - \Gamma^{\gamma}{}_{\mu\sigma}\Gamma^{2\beta}{}_{\gamma\nu} + \Gamma^{2\gamma}{}_{\nu\sigma}\Gamma^{\beta}{}_{\gamma\mu} - \Gamma^{2\gamma}{}_{\mu\sigma}\Gamma^{\beta}{}_{\gamma\nu} \qquad (4.60)$$

$$= ig^{\beta}{}_{\sigma}(B^1{}_{\nu\mu} + B^2{}_{\nu\mu}) + R^{1\beta}{}_{\sigma\nu\mu} + R^{2\beta}{}_{\sigma\nu\mu}$$

with

$$H^{\beta}{}_{\sigma\nu\mu} = \partial_\mu H^{\beta}{}_{\sigma\nu} - \partial_\nu H^{\beta}{}_{\sigma\mu} + H^{\gamma}{}_{\nu\sigma}H^{\beta}{}_{\gamma\mu} - H^{\gamma}{}_{\mu\sigma}H^{\beta}{}_{\gamma\nu} \qquad (4.61)$$
$$R^{1\beta}{}_{\sigma\nu\mu} = \partial_\mu\Gamma^{\beta}{}_{\sigma\nu} - \partial_\nu\Gamma^{\beta}{}_{\sigma\mu} + \Gamma^{\gamma}{}_{\nu\sigma}\Gamma^{\beta}{}_{\gamma\mu} - \Gamma^{\gamma}{}_{\mu\sigma}\Gamma^{\beta}{}_{\gamma\nu} \qquad (4.62)$$
$$R^{2\beta}{}_{\sigma\nu\mu\rho} = \partial_\mu\Gamma^{2\beta}{}_{\sigma\nu} - \partial_\nu\Gamma^{2\beta}{}_{\sigma\mu} + \Gamma^{2\gamma}{}_{\nu\sigma}\Gamma^{2\beta}{}_{\gamma\mu} - \Gamma^{2\gamma}{}_{\mu\sigma}\Gamma^{2\beta}{}_{\gamma\nu} +$$
$$+ \Gamma^{\gamma}{}_{\nu\sigma}\Gamma^{2\beta}{}_{\gamma\mu} - \Gamma^{\gamma}{}_{\mu\sigma}\Gamma^{2\beta}{}_{\gamma\nu} + \Gamma^{2\gamma}{}_{\nu\sigma}\Gamma^{\beta}{}_{\gamma\mu} - \Gamma^{2\gamma}{}_{\mu\sigma}\Gamma^{\beta}{}_{\gamma\nu} \qquad (4.63)$$

and where

$$F^1{}_{\kappa\mu} = \partial A^1{}_\mu/\partial x^\kappa - \partial A^1{}_\kappa/\partial x^\mu + i[A^1{}_\kappa, A^1{}_\mu] \qquad (4.64)$$
$$F_W{}^1{}_{\kappa\mu} = \partial W^1{}_\mu/\partial x^\kappa - \partial W^1{}_\kappa/\partial x^\mu + i[W^1{}_\kappa, W^1{}_\mu]$$
$$F_E{}^1{}_{\kappa\mu} = \partial A_E{}^1{}_\mu/\partial x^\kappa - \partial A_E{}^1{}_\kappa/\partial x^\mu$$
$$B^1{}_{\kappa\mu} = \partial B^1{}_\mu/\partial x^\kappa - \partial B^1{}_\kappa/\partial x^\mu + i[B^1{}_\kappa, B^1{}_\mu]$$

$$F^2{}_{\kappa\mu} = \partial A^2{}_\mu/\partial x^\kappa - \partial A^2{}_\kappa/\partial x^\mu + i[A^2{}_\kappa, A^2{}_\mu] + i[A^1{}_\kappa, A^2{}_\mu] + i[A^2{}_\kappa, A^1{}_\mu]$$
$$F_W{}^2{}_{\kappa\mu} = \partial W^2{}_\mu/\partial x^\kappa - \partial W^2{}_\kappa/\partial x^\mu + i[W^2{}_\kappa, W^2{}_\mu] + i[W^1{}_\kappa, W^2{}_\mu] + i[W^2{}_\kappa, W^1{}_\mu]$$
$$F_E{}^2{}_{\kappa\mu} = \partial A_E{}^2{}_\mu/\partial x^\kappa - \partial A_E{}^2{}_\kappa/\partial x^\mu$$
$$B^2{}_{\kappa\mu} = \partial B^2{}_\mu/\partial x^\kappa - \partial B^2{}_\kappa/\partial x^\mu + i[B^2{}_\mu, B^2{}_\kappa] + i[B^1{}_\mu, B^2{}_\kappa] + i[B^2{}_\mu, B^1{}_\kappa]$$

$R'^{\beta}{}_{\sigma\nu\mu}$ *is the Riemann-Christoffel curvature tensor for the complex 32-dimensional space generated by SU(3) and 4-dimensional General Coordinate transformations including secondary connections and metric with U(1)*⊗SU(2)⊗SU(3) *internal symmetry*.

Note $R'^{\beta}{}_{\sigma\nu\mu}$ factorizes into U(1)⊗SU(2)⊗SU(3) parts and a Riemann-Christoffel curvature tensor part due to the commutativity amongst $B^i{}_\mu$, $A^j{}_\mu$, $W^k{}_\mu$ and $A_E{}^k{}_\mu$ for all i, j, and k. For later use in defining a lagrangian we define

$$R'^{\beta}{}_{\sigma\nu\mu} = R'^1{}_E{}^{\beta}{}_{\sigma\nu\mu} + R'^2{}_E{}^{\beta}{}_{\sigma\nu\mu} + R'^1{}_{SU(2)}{}^{\beta}{}_{\sigma\nu\mu} + R'^2{}_{SU(2)}{}^{\beta}{}_{\sigma\nu\mu} + R'^1{}_{SU(3)}{}^{\beta}{}_{\sigma\nu\mu} + R'^2{}_{SU(3)}{}^{\beta}{}_{\sigma\nu\mu} +$$
$$+ R'^1{}_B{}^{\beta}{}_{\sigma\nu\mu} + R'^2{}_B{}^{\beta}{}_{\sigma\nu\mu} + R'^1{}_G{}^{\beta}{}_{\sigma\nu\mu} + R'^2{}_G{}^{\beta}{}_{\sigma\nu\mu} \qquad (4.65)$$

where

$$R'^{1}{}_{E}{}^{\beta}{}_{\sigma\nu\mu} = ig^{\beta}{}_{\sigma}F_{E}{}^{1}{}_{\nu\mu}$$
$$R'^{2}{}_{E}{}^{\beta}{}_{\sigma\nu\mu} = ig^{\beta}{}_{\sigma}F_{E}{}^{2}{}_{\nu\mu} \quad (4.66)$$

$$R'^{1}{}_{SU(2)}{}^{\beta}{}_{\sigma\nu\mu} = ig^{\beta}{}_{\sigma}F_{W}{}^{1}{}_{\nu\mu}$$
$$R'^{2}{}_{SU(2)}{}^{\beta}{}_{\sigma\nu\mu} = ig^{\beta}{}_{\sigma}F_{W}{}^{2}{}_{\nu\mu} \quad (4.67)$$

$$R'^{1}{}_{SU(3)}{}^{\beta}{}_{\sigma\nu\mu} = ig^{\beta}{}_{\sigma}F^{1}{}_{\nu\mu}$$
$$R'^{2}{}_{SU(3)}{}^{\beta}{}_{\sigma\nu\mu} = ig^{\beta}{}_{\sigma}F^{2}{}_{\nu\mu} \quad (4.68)$$

$$R'^{1}{}_{B}{}^{\beta}{}_{\sigma\nu\mu} = ig^{\beta}{}_{\sigma}B^{1}{}_{\nu\mu}$$
$$R'^{2}{}_{B}{}^{\beta}{}_{\sigma\nu\mu} = ig^{\beta}{}_{\sigma}B^{2}{}_{\nu\mu} \quad (4.69)$$

$$R'^{1}{}_{G}{}^{\beta}{}_{\sigma\nu\mu} = \partial_{\mu}\Gamma^{\beta}{}_{\sigma\nu} - \partial_{\nu}\Gamma^{\beta}{}_{\sigma\mu} + \Gamma^{\gamma}{}_{\nu\sigma}\Gamma^{\beta}{}_{\gamma\mu} - \Gamma^{\gamma}{}_{\mu\sigma}\Gamma^{\beta}{}_{\gamma\nu} \quad (4.70)$$
$$R'^{2}{}_{G}{}^{\beta}{}_{\sigma\nu\mu} = \partial_{\mu}\Gamma^{2\beta}{}_{\sigma\nu} - \partial_{\nu}\Gamma^{2\beta}{}_{\sigma\mu} + \Gamma^{2\gamma}{}_{\nu\sigma}\Gamma^{2\beta}{}_{\gamma\mu} - \Gamma^{2\gamma}{}_{\mu\sigma}\Gamma^{2\beta}{}_{\gamma\nu} +$$
$$+ \Gamma^{\gamma}{}_{\nu\sigma}\Gamma^{2\beta}{}_{\gamma\mu} - \Gamma^{\gamma}{}_{\mu\sigma}\Gamma^{2\beta}{}_{\gamma\nu} + \Gamma^{2\gamma}{}_{\nu\sigma}\Gamma^{\beta}{}_{\gamma\mu} - \Gamma^{2\gamma}{}_{\mu\sigma}\Gamma^{\beta}{}_{\gamma\nu}$$

The total Ricci tensor is

$$R'_{\sigma\mu} = R'^{\beta}{}_{\sigma\beta\mu}$$

$$= iF_{E}{}^{1}{}_{\sigma\mu} + iF_{E}{}^{2}{}_{\sigma\mu} + iF_{W}{}^{1}{}_{\sigma\mu} + iF_{W}{}^{2}{}_{\sigma\mu} + iF^{1}{}_{\sigma\mu} + iF^{2}{}_{\sigma\mu} + iB^{1}{}_{\sigma\mu} + iB^{2}{}_{\sigma\mu} +$$
$$+ \partial_{\mu}\Gamma^{\beta}{}_{\sigma\beta} - \partial_{\beta}\Gamma^{\beta}{}_{\sigma\mu} + \Gamma^{\gamma}{}_{\beta\sigma}\Gamma^{\beta}{}_{\gamma\mu} - \Gamma^{\gamma}{}_{\mu\sigma}\Gamma^{\beta}{}_{\gamma\beta} +$$
$$+ \partial_{\mu}\Gamma^{2\beta}{}_{\sigma\beta} - \partial_{\beta}\Gamma^{2\beta}{}_{\sigma\mu} + \Gamma^{2\gamma}{}_{\beta\sigma}\Gamma^{2\beta}{}_{\gamma\mu} - \Gamma^{2\gamma}{}_{\mu\sigma}\Gamma^{2\beta}{}_{\gamma\beta} + \Gamma^{\gamma}{}_{\beta\sigma}\Gamma^{2\beta}{}_{\gamma\mu} - \Gamma^{\gamma}{}_{\mu\sigma}\Gamma^{2\beta}{}_{\gamma\beta} + \Gamma^{2\gamma}{}_{\beta\sigma}\Gamma^{\beta}{}_{\gamma\mu} - \Gamma^{2\gamma}{}_{\mu\sigma}\Gamma^{\beta}{}_{\gamma\beta}$$

$$= R'^{1}{}_{E}{}^{\beta}{}_{\sigma\beta\mu} + R'^{2}{}_{E}{}^{\beta}{}_{\sigma\beta\mu} + R'^{1}{}_{SU(2)}{}^{\beta}{}_{\sigma\beta\mu} + R'^{2}{}_{SU(2)}{}^{\beta}{}_{\sigma\beta\mu} + R'^{1}{}_{SU(3)}{}^{\beta}{}_{\sigma\beta\mu} + R'^{2}{}_{SU(3)}{}^{\beta}{}_{\sigma\beta\mu} +$$
$$+ R'^{1}{}_{B}{}^{\beta}{}_{\sigma\beta\mu} + R'^{2}{}_{B}{}^{\beta}{}_{\sigma\beta\mu} + R'^{1}{}_{G}{}^{\beta}{}_{\sigma\beta\mu} + R'^{2}{}_{G}{}^{\beta}{}_{\sigma\beta\mu}$$
$$= R'^{1}{}_{E\sigma\mu} + R'^{2}{}_{E\sigma\mu} + R'^{1}{}_{SU(2)\sigma\mu} + R'^{2}{}_{SU(2)\sigma\mu} + R'^{1}{}_{SU(3)\sigma\mu} + R'^{2}{}_{SU(3)\sigma\mu} +$$
$$+ R'^{1}{}_{B}{}^{\beta}{}_{\sigma\beta\mu} + R'^{2}{}_{B}{}^{\beta}{}_{\sigma\beta\mu} + R'^{1}{}_{G\sigma\mu} + R'^{2}{}_{G\sigma\mu}$$
$$= R'^{1}{}_{\sigma\mu} + R'^{2}{}_{\sigma\mu} \quad (4.71)$$

where

$$R'^{1}{}_{\sigma\mu} = R'^{1}{}_{E\sigma\mu} + R'^{1}{}_{SU(2)\sigma\mu} + R'^{1}{}_{SU(3)\sigma\mu} + R'^{1}{}_{B}{}^{\beta}{}_{\sigma\beta\mu} + R'^{1}{}_{G\sigma\mu} \quad (4.72)$$
$$R'^{2}{}_{\sigma\mu} = R'^{2}{}_{E\sigma\mu} + R'^{2}{}_{SU(2)\sigma\mu} + R'^{2}{}_{SU(3)\sigma\mu} + R'^{2}{}_{B}{}^{\beta}{}_{\sigma\beta\mu} + R'^{2}{}_{G\sigma\mu}$$

$$R'^{1}{}_{E\sigma\mu} = iF_{E}{}^{1}{}_{\sigma\mu}$$
$$R'^{2}{}_{E\sigma\mu} = iF_{E}{}^{2}{}_{\sigma\mu}$$

$$R'^{1}{}_{SU(2)\sigma\mu} = iF_{W}{}^{1}{}_{\sigma\mu}$$

$$R'^{2}{}_{SU(2)\sigma\mu} = iF_{W}{}^{2}{}_{\sigma\mu}$$
$$R'^{1}{}_{SU(3)\sigma\mu} = iF^{1}{}_{\sigma\mu}$$
$$R'^{2}{}_{SU(3)\sigma\mu} = iF^{2}{}_{\sigma\mu}$$

$$R'^{1}{}_{B\sigma\mu} = iB^{1}{}_{\sigma\mu}$$
$$R'^{2}{}_{B\sigma\mu} = iB^{2}{}_{\sigma\mu}$$

with the further definition of $R''^{1}{}_{\sigma\mu}$ and $R''^{2}{}_{\sigma\mu}$:

$$R''^{1}{}_{\sigma\mu} = R'^{1}{}_{\sigma\mu} - R'^{1}{}_{E\sigma\mu} - R'^{1}{}_{W\sigma\mu} \qquad (4.73)$$
$$R''^{2}{}_{\sigma\mu} = R'^{2}{}_{\sigma\mu} - R'^{2}{}_{E\sigma\mu} - R'^{2}{}_{W\sigma\mu}$$

Eq. 4.71 is the Ricci tensor for the complex 32-dimensional space generated by SU(3) and 4-dimensional General Coordinate transformations when augmented with the secondary affine connections.

An additional Ricci-like tensor is

$$H_{\sigma\mu} = H^{\beta}{}_{\sigma\beta\mu} \qquad (4.74)$$

The curvature scalar is

$$R' = g^{\sigma\mu}R'_{\sigma\mu} = + \partial^{\sigma}\Gamma^{\beta}{}_{\sigma\beta} - \partial_{\beta}\Gamma^{\beta}{}_{\sigma}{}^{\sigma} + \Gamma^{\gamma}{}_{\beta\sigma}\Gamma^{\beta}{}_{\gamma}{}^{\sigma} - \Gamma^{\gamma}{}_{\mu\sigma}\Gamma^{\beta}{}_{\gamma\beta} + \partial^{\sigma}\Gamma^{2\beta}{}_{\sigma\beta} - \partial_{\beta}\Gamma^{2\beta}{}_{\sigma}{}^{\sigma} + \Gamma^{2\gamma}{}_{\beta\sigma}\Gamma^{2\beta}{}_{\gamma}{}^{\sigma} -$$
$$- \Gamma^{2\gamma\sigma}{}_{\sigma}\Gamma^{2\beta}{}_{\gamma\beta} + \Gamma^{\gamma}{}_{\beta\sigma}\Gamma^{2\beta}{}_{\gamma}{}^{\sigma} - \Gamma^{\gamma\sigma}{}_{\sigma}\Gamma^{2\beta}{}_{\gamma\beta} + \Gamma^{2\gamma}{}_{\beta\sigma}\Gamma^{\beta}{}_{\gamma}{}^{\sigma} - \Gamma^{2\gamma\sigma}{}_{\sigma}\Gamma^{\beta}{}_{\gamma\beta}$$

$$= g^{\sigma\mu}(R^{1\beta}{}_{\sigma\beta\mu} + R^{2\beta}{}_{\sigma\beta\mu}) \qquad (4.75)$$

Additional curvature scalars are

$$H = g^{\sigma\mu}H_{\sigma\mu} \qquad (4.76)$$
$$R'^{2} = g^{\sigma\mu}R'^{2}{}_{\sigma\mu} \qquad (4.77)$$

4.6 Separation of the Various Lagrangian Interaction Sectors

In this section we will separate the electromagnetic, Weak, gravitational and Strong interaction parts of the total lagrangian which we form as a generalization of the usual Einstein lagrangian with additional higher derivative terms added. The major aspect of our extension is the introduction of higher derivative terms in a such a manner that they can be handled by canonical lagrangian methods to obtain the dynamical equations of motion and the equal time commutation relations (for the 'free' field approximations.) We also separate the Ricci tensor into two parts in order to use pseudoQuantization field theory (with fields labeled '1' and '2') to implement canonical lagrangian methods, and to introduce the flat space metric $\eta^{\sigma\mu}$ by a Higgs

Mechanism.[25] The constant flat space metric is usually an assumed quantity. But its close relation to the quantum field $g^{\sigma\mu}$ suggests that it could be generated by the same Higgs Mechanism that generates particle mass constants.[26]

Therefore we assume the lagrangian density:[27]

$$\mathcal{L} = \text{Tr}\,\sqrt{g}[MD_\nu R''^{1}{}_{\sigma\mu}D^\nu R''^{2\sigma\mu} + aR'^{1}{}_{\sigma\mu}R'^{2\sigma\mu} + bR' + cg^{\sigma\mu}g^{2}{}_{\sigma\mu} + c'g^{2\sigma\mu}g^{2}{}_{\sigma\mu} - dA^{2}{}_{\mu}A^{2\mu}] \qquad (4.78)$$

where M, a, b, c, c', and d are constants to be determined later.

This higher derivative lagrangian maintains the locality of the theory but does entail a modest modification in the derivation of the Euler-Lagrange equations of motion. It also requires the use of principal value propagators rather than ordinary Feynman propagators for gluon and graviton interactions. Thus the Electromagnetic sector, the Strong Interaction sector and the Gravitation sector are Action-at-a-Distance theories that are similar in spirit to Wheeler-Feynman Electrodynamics. The Weak sector may, or may not, be an Action-at-a-Distance theory. It is not constrained to be Action-at-a-Distance by the present considerations.

Since we wish to apply it cosmologically, and within hadrons, where the gravitational spinor connections are negligible due to the smallness of the gravitational constant G and the 'smallness' of B spin on the cosmological scale, we set $B^{1}{}_{\nu\mu} = B^{2}{}_{\nu\mu} = 0$ and find[28]

$$\begin{aligned}\mathcal{L} = \text{Tr}\,\sqrt{g}[&MD_\nu(R'^{1}{}_{SU(3)\sigma\mu} + R'^{1}{}_{G\sigma\mu})D^\nu(R'^{2}{}_{SU(3)}{}^{\sigma\mu} + R'^{2}{}_{G}{}^{\sigma\mu}) + \\ &+ a(R'^{1}{}_{E\sigma\mu} + R'^{1}{}_{SU(2)\sigma\mu} + R'^{1}{}_{SU(3)\sigma\mu} + R'^{1}{}_{G\sigma\mu})(R'^{2}{}_{E}{}^{\sigma\mu} + R'^{2}{}_{SU(2)}{}^{\sigma\mu} + R'^{2}{}_{SU(3)}{}^{\sigma\mu} + R'^{2}{}_{G}{}^{\sigma\mu}) + \\ &+ bR' + cg^{\sigma\mu}g^{2}{}_{\sigma\mu} + c'g^{2\sigma\mu}g^{2}{}_{\sigma\mu} - dA^{2}{}_{\mu}A^{2\mu}]\end{aligned}$$

$$\begin{aligned}= \text{Tr}\,\sqrt{g}[&M\{D_\nu R'^{1}{}_{SU(3)\sigma\mu}D^\nu R'^{2}{}_{SU(3)}{}^{\sigma\mu} + D_\nu R'^{1}{}_{G\sigma\mu}D^\nu R'^{2}{}_{G}{}^{\sigma\mu} + \\ &+ D_\nu R'^{1}{}_{G\sigma\mu}D^\nu R'^{2}{}_{SU(3)}{}^{\sigma\mu} + D_\nu R'^{1}{}_{SU(3)\sigma\mu}D^\nu R'^{2}{}_{G}{}^{\sigma\mu}\} + \\ &+ a\{(R'^{1}{}_{E\sigma\mu} + R'^{1}{}_{SU(2)\sigma\mu} + R'^{1}{}_{SU(3)\sigma\mu})(R'^{2}{}_{E}{}^{\sigma\mu} + R'^{2}{}_{SU(2)}{}^{\sigma\mu} + R'^{2}{}_{SU(3)}{}^{\sigma\mu}) + R'^{1}{}_{G\sigma\mu}R'^{2}{}_{G}{}^{\sigma\mu} + \\ &+ R'^{1}{}_{G\sigma\mu}(R'^{2}{}_{E}{}^{\sigma\mu} + R'^{2}{}_{SU(2)}{}^{\sigma\mu} + R'^{2}{}_{SU(3)}{}^{\sigma\mu}) + (R'^{1}{}_{E\sigma\mu} + R'^{1}{}_{SU(2)\sigma\mu} + R'^{1}{}_{SU(3)\sigma\mu})R'^{2}{}_{G}{}^{\sigma\mu}\} + \\ &+ bg^{\sigma\mu}(R'^{1}{}_{G}{}^{\beta}{}_{\sigma\beta\mu} + R'^{2}{}_{G}{}^{\beta}{}_{\sigma\beta\mu}) + cg^{\sigma\mu}g^{2}{}_{\sigma\mu} + c'g^{2\sigma\mu}g^{2}{}_{\sigma\mu} - dA^{2}{}_{\mu}A^{2\mu}]\end{aligned} \qquad (4.79)$$

Since there are no strong interaction fields in 'empty' space and gravity is negligible within hadrons,[29] we can drop the interaction terms between these interactions. However we cannot drop the interaction terms amongst Electromagnetism, the Weak interaction, and the

---

[25] In earlier books such as Blaha (2016f) we showed that the use of two fields for each particle type enables us to clearly separate the 'vacuum expectation value' from its associated second quantized 'Higgs' field. The application to the weak field approximation for gravitons is one example.

[26] See also Blaha (2016c).

[27] Since the lagrangian terms are matrices it is necessary to take the trace.

[28] The constants in eq. 4.62 have the dimensions: M has the dimension of inverse mass squared, b has dimension mass squared, a is dimensionless, c and c' have dimension mass to the 4th order, and d has dimension mass squared.

[29] We show gravity weakens at very short distances using our Two-Tier Quantum Field Theory formalism. See Blaha (2003) and (2005a) among other books by the author.

Strong Interaction within, and between, hadrons, and the interaction terms between Electromagnetism and Gravitation cosmologically.

Eq. 4.79 can therefore be expressed as:[30]

$$\mathcal{L} = \mathcal{L}_E + \mathcal{L}_{SU(2)} + \mathcal{L}_{SU(3)} + \mathcal{L}_G + \mathcal{L}_{int} \tag{4.80}$$

where taking traces of $\mathcal{L}$s terms is understood[31]

$$\begin{aligned}\mathcal{L}_E &= \mathrm{Tr}\,\sqrt{g}\{M\{[\partial_\nu + i(A_E{}^1{}_\nu + A_E{}^2{}_\nu)]F^1{}_{E\sigma\mu}[\partial^\nu + i(A_E{}^{1\nu} + A_E{}^{2\nu})]F^2{}_E{}^{\sigma\mu}\} + aF^1{}_{E\sigma\mu}F^2{}_E{}^{\sigma\mu}\} \\ \mathcal{L}_{SU(2)} &= \mathrm{Tr}\,\sqrt{g}[aF_W{}^1{}_{\sigma\mu}F_W{}^{2\sigma\mu}] \\ \mathcal{L}_{SU(3)} &= \mathrm{Tr}\,\sqrt{g}\{M[\partial_\nu + i(A^1{}_\nu + A^2{}_\nu)]F^1{}_{\sigma\mu}[\partial^\nu + i(A^{1\nu} + A^{2\nu})]F^{2\sigma\mu} + aF^1{}_{\sigma\mu}F^{2\sigma\mu} - dA^2{}_\mu A^{2\mu}\}\end{aligned} \tag{4.81}$$

$$\begin{aligned}\mathcal{L}_G &= \mathrm{Tr}\,\sqrt{g}[MD_\nu R'^1{}_{G\sigma\mu}D^\nu R'^2{}_G{}^{\sigma\mu} + aR'^1{}_{G\sigma\mu}R'^2{}_G{}^{\sigma\mu} + bg^{\sigma\mu}(R'^1{}_G{}^\beta{}_{\sigma\beta\mu} + R'^2{}_G{}^\beta{}_{\sigma\beta\mu}) + cg^{\sigma\mu}g^2{}_{\sigma\mu} + c'g^{2\sigma\mu}g^2{}_{\sigma\mu}] \\ &= \mathrm{Tr}\,\sqrt{g}[MD_\nu R'^1{}_{G\sigma\mu}D^\nu R'^2{}_G{}^{\sigma\mu} + aR'^1{}_{G\sigma\mu}R'^2{}_G{}^{\sigma\mu} + bH + cg^{\sigma\mu}g^2{}_{\sigma\mu} + c'g^{2\sigma\mu}g^2{}_{\sigma\mu}]\end{aligned}$$

$$\begin{aligned}\mathcal{L}_{int} = \mathrm{Tr}\,\sqrt{g}[M\{&-i(A_E{}^1{}_\nu + A_E{}^2{}_\nu + W^1{}_\nu + W^2{}_\nu)F^1{}_{\sigma\mu}D^\nu F^{2\sigma\mu} - iD_\nu F^1{}_{\sigma\mu}(A_E{}^{1\nu} + A_E{}^{2\nu} + \\ &+ W^{1\nu} + W^{2\nu})F^{2\sigma\mu} + iD_\nu R'^1{}_{G\sigma\mu}[\partial^\nu + i(A^{1\nu} + A^{2\nu} + A_E{}^{1\nu} + A_E{}^{2\nu} + W^{1\nu} + W^{2\nu})]F^{2\sigma\mu} + \\ &+ i[\partial_\nu + i(A^1{}_\nu + A^2{}_\nu + A_E{}^1{}_\nu + A_E{}^2{}_\nu + + W^1{}_\nu + W^2{}_\nu)]F^1{}_{\sigma\mu}D^\nu R'^2{}_G{}^{\sigma\mu}\} + \\ &+ a\{-F_E{}^1{}_{\sigma\mu}(F_W{}^{2\sigma\mu} + F^{2\sigma\mu}) - F^1{}_{\sigma\mu}(F_E{}^{2\sigma\mu} + F_W{}^{2\sigma\mu}) - F_W{}^1{}_{\sigma\mu}(F_E{}^{2\sigma\mu} + F^{2\sigma\mu}) + \\ &+ iR'^1{}_{G\sigma\mu}(F_E{}^{2\sigma\mu} + F_W{}^{2\sigma\mu} + F^{2\sigma\mu}) + i(F_E{}^1{}_{\sigma\mu} + F_W{}^1{}_{\sigma\mu} + F^1{}_{\sigma\mu})R'^2{}_G{}^{\sigma\mu}\}]\end{aligned} \tag{4.82}$$

Thus $\mathcal{L}_{SU(3)}$, $\mathcal{L}_{SU(2)}$, $\mathcal{L}_E$ and $\mathcal{L}_{int}$ are the dominant interactions within hadrons, and $\mathcal{L}_G$, $\mathcal{L}_E$ and $\mathcal{L}_{int}$ are the dominant interactions in space within the framework of this discussion.

The $D_\nu R'^1{}_{G\sigma\mu}$ and $D^\nu R'^2{}_G{}^{\sigma\mu}$ terms have the form:

$$D_\nu R'^i{}_{G\sigma\mu} = + \partial_\nu R'^i{}_{G\sigma\mu} \;\; - \Gamma^\beta{}_{\sigma\nu}R'^i{}_{G\beta\mu} - \Gamma^\beta{}_{\nu\mu}R'^i{}_{G\sigma\beta}$$

for i = 1, 2.

One may ask why we relate these four interactions: they all have coupling constants with comparable values when their coupling constants are generated by the 'Higgs' Mechanism. Further the basis of the unification in the four interaction Riemann-Christoffel curvature tensor leads to interactions beyond those in The Standard Model. These additional interactions in $\mathcal{L}_{int}$ *lead to new phenomena in this unified theory such as: 1) a possible relationship between the coupling constants appearing below; 2) a possible relationship between parts of these interactions, 3) a possible solution of the proton spin puzzle due to electromagnetic-gluon terms in* $\mathcal{L}_{int}$ *that have not been previously considered, and 4) a possible*

[30] We only consider the gauge field lagrangian terms in this chapter and in this book.

[31] The coupling constants of the gauge fields are not shown in the interests of simplicity. See eq. 4.83 for the coupling constants, which will be treated as implicit in the gauge fields in the remainder of this chapter.

explanation of the proton radius puzzle. These possibilities will be discussed in subsequent chapters.

We now introduce a strong interaction coupling constant f, an electromagnetic coupling constant e, and a Weak coupling constant g_W with[32]

$$\begin{aligned} A^1_{\ \mu} &\rightarrow fA^{1\mu} \\ A^2_{\ \mu} &\rightarrow fA^{2\mu} \\ A_E{}^1_{\ \mu} &\rightarrow eA_E{}^{1\mu} \\ A_E{}^2_{\ \mu} &\rightarrow eA_E{}^{2\mu} \\ W^1_{\ \mu} &\rightarrow g_W W^{1\mu} \\ W^2_{\ \mu} &\rightarrow g_W W^{2\mu} \end{aligned} \tag{4.83}$$

leading to

$$\begin{aligned} F^1_{\ \kappa\mu} &= \partial A^1_{\ \mu}/\partial x^\kappa - \partial A^1_{\ \kappa}/\partial x^\mu + if[A^1_{\ \kappa}, A^1_{\ \mu}] \\ F^2_{\ \kappa\mu} &= \partial A^2_{\ \mu}/\partial x^\kappa - \partial A^2_{\ \kappa}/\partial x^\mu + if[A^2_{\ \kappa}, A^2_{\ \mu}] + if[A^1_{\ \kappa}, A^2_{\ \mu}] + if[A^2_{\ \kappa}, A^1_{\ \mu}] \\ F_W{}^1_{\ \kappa\mu} &= \partial W^1_{\ \mu}/\partial x^\kappa - \partial W^1_{\ \kappa}/\partial x^\mu + ig_W[W^1_{\ \kappa}, W^1_{\ \mu}] \\ F_W{}^2_{\ \kappa\mu} &= \partial W^2_{\ \mu}/\partial x^\kappa - \partial W^2_{\ \kappa}/\partial x^\mu + ig_W[W^2_{\ \kappa}, W^2_{\ \mu}] + ig_W[W^1_{\ \kappa}, W^2_{\ \mu}] + ig_W[W^2_{\ \kappa}, W^1_{\ \mu}] \end{aligned} \tag{4.84}$$

Then the Strong Interaction lagrangian density terms are modified to:[33,34]

$$\mathcal{L}_{SU(3)} = \text{Tr}\,\sqrt{g}[MD_{SU(3)\nu}R'^1{}_{SU(3)\sigma\mu}D_{SU(3)}{}^\nu R'^2{}_{SU(3)}{}^{\sigma\mu} + aR'^1{}_{SU(3)\sigma\mu}R'^2{}_{SU(3)}{}^{\sigma\mu} - dA^2_{\ \mu}A^{2\mu}] \tag{4.85}$$

with

$$D_{SU(3)\nu} = [\partial_\nu + if(A^1_{\ \nu} + A^2_{\ \nu})]$$

and the electromagnetic lagrangian density term is now

$$\mathcal{L}_E = \sqrt{g}\{aF^1{}_{E\sigma\mu}F^2{}_E{}^{\sigma\mu}\} \tag{4.86}$$

Corresponding changes take place in $\mathcal{L}_{SU(2)}$ and $\mathcal{L}_{int}$.

We now approximate the metric determinant as g = 1 within hadrons. Thus the Strong lagrangian part becomes

$$\begin{aligned} \mathcal{L}_{SU(3)} &= \text{Tr}\,\{MD_\nu R'^1{}_{SU(3)\sigma\mu}D^\nu R'^2{}_{SU(3)}{}^{\sigma\mu} + aR'^1{}_{SU(3)\sigma\mu}R'^2{}_{SU(3)}{}^{\sigma\mu} - dA^2_{\ \mu}A^{2\mu}\} && (4.87) \\ &= \text{Tr}\,\{MD_\nu R'^1{}_{SU(3)\sigma\mu}D^\nu R'^2{}_{SU(3)}{}^{\sigma\mu} + \varsigma R'^1{}_{SU(3)\sigma\mu}R'^2{}_{SU(3)}{}^{\sigma\mu} - \varsigma A^2_{\ \mu}A^{2\mu}\} && (4.88) \end{aligned}$$

where

---

[32] The coupling constants are assumed to be the renormalized physical coupling constants.

[33] The form is virtually identical to S. Blaha, Phys. Rev. D**11**, 2921 (1974), and Blaha's 1976 Gravity Research Foundation Essay (Honorable Mention), except for the initial derivative term.

[34] We note the constant a, which appears in this chapter and chapter 5 is NOT the Charmonium constant a in eq. 2.1.

$$D_\nu R'^{j}{}_{SU(3)\sigma\mu} = \partial_\nu R'^{j}{}_{SU(3)\sigma\mu} + f[A^1{}_\nu, R'^{j}{}_{SU(3)\sigma\mu}] \qquad (4.89)$$

for j = 1, 2, and where

$$\zeta = a = -\tfrac{1}{2} \qquad (4.90)$$

$$\varsigma = d = \tfrac{1}{2}\lambda^2 \qquad (4.91)$$

We note that we will choose

$$b = (16\pi G)^{-1} \qquad (4.92)$$

in accord with the usual Einstein dynamic equations. We will use the experimentally determined value for the electromagnetic coupling constant $e = (4\pi\alpha)^{1/2}$.

5. SU(3) Strong Interaction Dynamics

5.1 16-Dimensional Complex-Valued Coordinates

Earlier in section 4.4 we showed that the 32-dimensional complex-valued space corresponding to the vierbein $l^{\mu ai}(x)$ described in section 4.2 can be mapped to a 16-dimensional complex-valued space. This space can contain our 4-dimensional universe with complex-valued coordinates due to the well-known map of our universe into complex 10-dimensional space described in Eddingon (1952). Since we chose, in earlier books, to take our universe to have complex-valued coordinates the 'Eddington' map requires a 10-dimensional complex-valued space. Thus the enclosing 16-dimensional complex universe, that we have called the Megaverse, easily contains our universe, and probably other universes, as surfaces within it.

Since our universe has complex-valued coordinates[35] we require complex-valued arguments for the Strong Interaction gauge fields that are consistent with the form of the coordinates of quarks with which they interact:

$$\begin{array}{l}\text{Real-valued Energies} \\ \text{Complex-valued spatial Momenta: } \mathbf{x} = \mathbf{x}_r + i\mathbf{x}_i\end{array} \tag{5.1}$$

This form of the coordinates is implemented after the extraction of the dynamical equations from the lagrangian.

On the other hand, due to the effective factorization of the lagrangian into separate electromagnetic, gravitational, Weak and Strong Interaction parts we can use complex-valued coordinates in the gravitational sector without regard to the restrictions of eq. 5.1 on fermions.

5.2 Strong Interaction Lagragian Terms

The Strong Interaction lagrangian (eq. 4.69) with color fermion terms added is[36]

$$\mathcal{L}_{SU(3)} = \text{Tr}\ \{MD_\nu F^1{}_{\sigma\mu}D^\nu F^{2\sigma\mu} + aF^1{}_{\sigma\mu}F^{2\sigma\mu} - dA^2{}_\mu A^{2\mu}\} + \bar{\psi}[i\not\nabla + f(\not A^1 + \not A^2) - m]\psi \tag{5.2}$$

where (for j = 1, 2)

$$D_\nu F^j{}_{\sigma\mu} = \partial_\nu F^j{}_{\sigma\mu} + f[A^1{}_\nu, F^j{}_{\sigma\mu}] \tag{5.2a}$$

---

[35] In earlier work we showed that the four known species of fermions: charged leptons, neutral leptons, up-type quarks and down-type quarks, require complex space-time coordinates and the complex Lorentz group. In particular, quarks of both species require real-valued energies and complex-valued spatial momenta of the form $\mathbf{x} = \mathbf{x}_r + i\mathbf{x}_i$. We call particles of this type *complexons*.

[36] We note the constant a, that appears in this chapter and chapter 4, is NOT the Charmonium constant a in eq. 2.1.

The lagrangian is equal to eq. 17 of S. Blaha, Phys. Rev. **D11**, 2921 (1974) except for additional terms $MD_\nu F^1{}_{\sigma\mu}D^\nu F^{2\sigma\mu}$ and $[A^2{}_\kappa, A^2{}_\mu]$; and the following changes in parameters:

$$a = -\tfrac{1}{2} \qquad\qquad d = \tfrac{1}{2}\lambda^2$$

Since the paper essentially contains a complete description of our Strong Interaction theory (modulo the additional terms) we refer the reader to it and to its predecessor paper referenced therein. There are a few additional changes required to bring the 1974 papers into agreement with our current theory:

1. Eq. 30 of the above referenced paper must be modified to

$$F^2{}_{\kappa\mu} = \partial A^2{}_\mu/\partial x^\kappa - \partial A^2{}_\kappa/\partial x^\mu + \mathbf{if[A^2{}_\kappa, A^2{}_\mu]} + if[A^1{}_\kappa, A^2{}_\mu] + if[A^2{}_\kappa, A^1{}_\mu] \qquad (5.3)$$

with the addition of the 'bolded' term $if[A^2{}_\kappa, A^2{}_\mu]$. There is also a trivial change of notation of coupling constant from 'g' to 'f'.

2. Eqs. 6 and 18 should have the interaction term expanded to

$$g\mathcal{A}^1 \rightarrow g(\mathcal{A}^1 + \mathcal{A}^2) \qquad \ldots \qquad (5.4)$$

and similarly in eq. 20. Eqs. 38 – 41 directly show that the additional interaction term leads to a gluon propagator[37] $<A^1 + A^2, A^1 + A^2> = 2<A^1, A^2> + <A^1, A^1>$, and introduces a 1/r term in the potential part of the gluon propagator.

As a result the effective gluon propagator in the theory, **if the $Mf^2D_\nu F^1{}_{\sigma\mu}D^\nu F^{2\sigma\mu}$ term is neglected**, combines eqs. 38 and 39 to give the *short-distance*[38] gluon propagator between quarks:

$$g_{\mu\nu}\delta_{ab}P[\lambda^2/k^4 - 1/k^2] \qquad (5.5)$$

up to a constant factor.[39]

This, in turn, *explicitly* leads to a Strong Interaction potential of the form

$$V(r) = -2f^2/r + f^2\lambda^2 r \qquad (5.6)$$

[37] Eqs. 40-41 in the above referenced paper.

[38] We anticipate that the $Mf^2D_\nu F^1{}_{\sigma\mu}D^\nu F^{2\sigma\mu}$ term will affect the short-distance behavior of the inter-quark interaction. The equivalent term in the gravitation sector influences the long-distance form of the gravity potential and leads to a MoND-like behavior. See chapter 6.

[39] This propagator is taken in Principal value to avoid potential unitarity problems. This topic is described in detail in earlier papers and books by the author.

Naturally one can expect perturbative corrections to eq. 5.6 in higher order in f. However, as will be seen later, the apparent relative smallness of f suggests eq. 5.6 is a good approximation to the *short-distance*, inter-quark interaction.

5.2.1 Canonical Equal Time Commutation Relations

The Euler-Lagrange equations of motion, eqs. 27 – 31 in the 1974 paper, are modified most significantly by the $Mf^2D_\nu F^1{}_{\sigma\mu}D^\nu F^{2\sigma\mu}$ lagrangian term in eq. 5.2. In order to follow the canonical method to obtain the contributions to the equations of motion of this higher derivative term we will use integration by parts and discard surface terms, as is usually done in quantum field theory. We should start with eq. 5.2. However, with a view towards perturbation theory which appears reasonable in view of the smallness of the strong interaction coupling constant $f^2/4\pi = 0.024$ seen below, we will abstract a quadratic expresssion in the fields from eq. 5.2 and then proceed to develop gluon propagators and the strong interaction potential. The 'free' Strong Interaction lagrangian that we use is

$$\mathcal{L}_{SU(3)F} = \text{Tr}\{MD_{F\nu}F_F{}^{1a}{}_{\sigma\mu}D_F{}^\nu F_F{}^{2a\sigma\mu} + aF_F{}^{1a}{}_{\sigma\mu}F_F{}^{2a\sigma\mu} - dA^{2a}{}_\mu A^{2a\mu}\} + \bar{\psi}[i\not{\nabla} + f(\not{A}^1 + \not{A}^2) - m]\psi \tag{5.7}$$

where

$$F^{1a}{}_{\mu\kappa} = \partial A^{1a}{}_\mu/\partial x^\kappa - \partial A^{1a}{}_\kappa/\partial x^\mu$$
$$F^{2a}{}_{\mu\kappa} = \partial A^{2a}{}_\mu/\partial x^\kappa - \partial A^{2a}{}_\kappa/\partial x^\mu \tag{5.8}$$

and

$$D_{F\nu} = \partial_\nu \tag{5.9}$$

The conjugate momenta to $A^{1a}{}_\mu$ and $A^{2a}{}_\mu$ are respectively

$$\pi^{1a}{}_\mu = \partial\mathcal{L}_{SU(3)F}/(\partial A^{1a}{}_\mu/\partial t) = aF_F{}^{2a\mu t}$$
$$\pi^{2a}{}_\mu = \partial\mathcal{L}_{SU(3)F}/(\partial A^{2a}{}_\mu/\partial t) = aF_F{}^{1a\mu t} \tag{5.10}$$

The non-zero, equal time commutation relations are

$$[\pi^{ia}{}_\mu(\mathbf{x}, t), A^{jb}{}_\nu(\mathbf{y}, t)] = i(1 - \delta^{ij})\delta^{ab}\delta^{G(\mu\nu)}(\mathbf{x} - \mathbf{y}) \tag{5.11}$$

where i and j label the fields, and $G(\mu\nu)$ indicates the gauge[40] G and the associated index expressions, with

$$\delta^{G(\mu\nu)}(\mathbf{x} - \mathbf{y}) = \int d^4k \exp(-ik\cdot x)b^G{}_{\mu\nu}(k)/(2\pi)^4 \tag{5.12}$$

where $b^G{}_{\mu\nu}(k)$ is a polynomial in k with a δ-function factor restricting the integration over k.

---

[40] Not the gravitational coupling constant.

5.2.2 Dynamical Equations

After performing partial integrations on the $MD_{F\nu}F_F{}^{1}{}_{\sigma\mu}D_F{}^{\nu}F_F{}^{2\sigma\mu}$ term (and discarding surface terms at 'infinity') the Euler-Lagrange dynamical equations (in the Landau gauge) due to independent variations with respect to $A^{1a}{}_{\mu}$ is

$$2M\Box^2 A^{2a}{}_{\mu} - 2a\Box A^{2a}{}_{\mu} = -f\bar{\psi}T^a\gamma_{\mu}\psi \tag{5.13}$$

And, with respect to $A^{2a}{}_{\mu}$, is

$$2M\Box^2 A^{1a}{}_{\mu} - 2a\Box A^{1a}{}_{\mu} - 2dA^{2a}{}_{\mu} = -f\bar{\psi}T^a\gamma_{\mu}\psi \tag{5.14}$$

where T^a is an SU(3) generator. Subtracting the equations we find

$$2M\Box^2 A^{1a}{}_{\mu} - 2a\Box A^{1a}{}_{\mu} - 2M\Box^2 A^{2a}{}_{\mu} + 2a\Box A^{2a}{}_{\mu} - 2dA^{2a}{}_{\mu} = 0$$

or

$$A^{2a}{}_{\mu} = [2M\Box^2 - 2a\Box + 2d]^{-1}[2M\Box^2 A^{1a}{}_{\mu} - 2a\Box A^{1a}{}_{\mu}] \tag{5.15}$$

with the result

$$\{2M\Box^2 - 2a\Box - 2d[2M\Box^2 - 2a\Box + 2d]^{-1}[2M\Box^2 - 2a\Box]\}A^{1a}{}_{\mu} = -f\bar{\psi}T^a\gamma_{\mu}\psi$$

or

$$\{2M\Box^2 - 2a\Box - 2d[2M\Box^2 - 2a\Box + 2d]^{-1}[2M\Box^2 - 2a\Box]\}A^{1a}{}_{\mu} = -f\bar{\psi}T^a\gamma_{\mu}\psi$$

$$\{2M\Box^2 - 2a\Box - 2d + 4d^2[2M\Box^2 - 2a\Box + 2d]^{-1}\}A^{1a}{}_{\mu} = -f\bar{\psi}T^a\gamma_{\mu}\psi \tag{5.16}$$

Eq. 5.16 leads to the Principal Value (Feynman) propagator:

$$\begin{aligned} D^{11}{}_{\mu\nu}(x-y) &= P -i\langle 0|T(A^1{}_{\mu}(x), A^1{}_{\nu}(y)|0\rangle \\ &= P\int d^4k\, \exp(-ik\cdot x) b_{\mu\nu}(k) D_1(k)/(2\pi)^4 \end{aligned} \tag{5.17}$$

where $b_{\mu\nu}(k)$ is a Landau gauge polynomial in k, and

$$\begin{aligned} D_1(k) &= \{2Mk^4 - 2ak^2 - 2d + 4d^2[2Mk^4 - 2ak^2 + 2d]^{-1}\}^{-1} \\ &= [2Mk^4 - 2ak^2 + 2d](2Mk^4 - 2ak^2)^{-2} \end{aligned}$$

Thus

$$\begin{aligned} D^{11}{}_{\mu\nu}(x-y) &= P\int d^4k\, \exp(-ik\cdot x) b_{\mu\nu}(k)[2Mk^4 - 2ak^2 + 2d]/[(2\pi)^4(2Mk^4 - 2ak^2)^2] \\ &= P\int d^4k\, \exp(-ik\cdot x) b_{\mu\nu}(k)[2Mk^4 - 2ak^2 + 2d]/[(2\pi)^4 k^4(2Mk^2 - 2a)^2] \end{aligned} \tag{5.18}$$

indicating a linear potential r term as well as terms of lower powers in r and Yukawa-like terms with a mass of $(a/M)^{1/2}$. We will describe the resulting effective Strong Interaction potential in more detail later.

Eq. 5.11 leads to the other propagator:

$$D^{12}{}_{\mu\nu}(x-y) = P\, -i\langle 0|T(A^1{}_\mu(x), A^2{}_\nu(y)|0\rangle$$
$$= P\int d^4k\, \exp(-ik\cdot x) b_{\mu\nu}(k) D_2(k)/(2\pi)^4 \quad (5.19)$$

where

$$D_2(k) = [2Mk^4 - 2ak^2]^{-1} \quad (5.20)$$

Thus

$$D^{12}{}_{\mu\nu}(x-y) = P\int d^4k\, \exp(-ik\cdot x) b_{\mu\nu}(k)/[(2\pi)^4 k^2(2Mk^2 - 2a)] \quad (5.21)$$

indicating a 1/r potential term plus a Yukawa term with a mass of $(a/M)^{1/2}$.

Due to the form of the interaction with quarks (eq. 5.7) the total effective gluon interaction between quarks is

$$D^{tot}{}_{\mu\nu}(x-y) = D^{11}{}_{\mu\nu}(x-y) + 2D^{12}{}_{\mu\nu}(x-y)$$
$$= P\int d^4k\, \exp(-ik\cdot x) b_{\mu\nu}(k)\{[3Mk^4 - 3ak^2 + 2d]/[(2\pi)^4 k^4(2Mk^2 - 2a)^2]\} \quad (5.22)$$

5.2.2 Strong Interaction Potential

Eq. 5.22 leads to the form of the total Strong Interaction potential. We note that the $\mu = \nu = 0$ part of the Feynman propagator for transverse gluons has the form:

$$D^{tot}{}_{00}(x-y) = \ldots - \int d^4k\, V_{SI}(\mathbf{k}) \exp(-ik\cdot(x-y))/(2\pi)^4 = \ldots + V_{SI}(\mathbf{x}-\mathbf{y})\delta(x_0 - y_0) \quad (5.23)$$

where

$$V_{SI}(\mathbf{x}) = -\int d^3k\, \exp(i\mathbf{k}\cdot\mathbf{x}) V_{SI}(\mathbf{k})/(2\pi)^3$$

$$V_{SI}(\mathbf{k}) = [3M\mathbf{k}^4 + 3a\mathbf{k}^2 + 2d]/[\mathbf{k}^4(2M\mathbf{k}^2 + 2a)^2] \quad (5.24)$$
$$= (2M)^{-2}\{2d(M/a)^2/\mathbf{k}^4 + [3a(M/a)^2 - 4d(M/a)^3]/\mathbf{k}^2$$
$$+ 2d(M/a)^2/(\mathbf{k}^2 + a/M)^2 + [-3a(M/a)^2 + 4d(M/a)^3]/(\mathbf{k}^2 + a/M)\}$$

Letting

$$m_{SI} = (a/M)^{1/2} \quad (5.25)$$

we find

$$V_{SI}(\mathbf{k}) = (2a)^{-2}\{2d/\mathbf{k}^4 + [3a - 4dm_{SI}{}^{-2}]/\mathbf{k}^2 + 2d/(\mathbf{k}^2 + m_{SI}{}^2)^2 + [4dm_{SI}{}^{-2} - 3a]/(\mathbf{k}^2 + m_{SI}{}^2)\} \quad (5.26)$$

The constant, a, is dimensionless and of order 1. The constant M has the dimension of inverse mass squared. We anticipate M will be extremely large resulting in a very small gluon mass m_{SI}.

There are massless gluon terms that generate color confinement reducing the impact of the massive gluon terms to a negligible effect outside hadronic regions.

We also see that the value of the inverse of graviton masses is of the order of the average galactic radius (the average galactic radius is large) and thus generate a Modified Newtonian potential (MoND) as will be seen in chapter 6.

Substituting eq. 5.26 we obtain a sum of massless and Yukawa-like potentials. The Yukawa potential is

$$V_Y(\mathbf{r}) = \int d^3k \exp(i\mathbf{k}\cdot\mathbf{r})/[(2\pi)^3(\mathbf{k}^2 + m^2)] = \exp(-mr)/[4\pi r] \qquad (5.27)$$

Thus we obtain

$$V_{SI}(\mathbf{r}) = -(2a)^{-2}\{2d(dV_Y(\mathbf{r})/dm^2)|_{m=0} + [3a - 4dm_{SI}^{-2}]/(4\pi r) - 2d(dV_Y(\mathbf{r})/dm^2|_{m=m_{SI}}) + [4dm_{SI}^{-2} - 3a]V_Y(\mathbf{r})|_{m=m_{SI}}\} \qquad (5.28)$$

with the form

$$V_{SI}(\mathbf{r}) = \alpha_1 r + \alpha_2/r + \alpha_3 e^{-m_{SI}r}/(4\pi m_{SI}) + \alpha_4 e^{-m_{SI}r}/(4\pi r) \qquad (5.29)$$

where the constants α_i are:

$$\alpha_1 = -(2a)^{-2}d/(8\pi) \quad \text{(up to an infrared divergent constant)} \qquad (5.30)$$
$$\alpha_2 = -(2a)^{-2}[3a - 4dm_{SI}^{-2}]/(4\pi)$$
$$\alpha_3 = -(2a)^{-2}d$$
$$\alpha_4 = -(2a)^{-2}[4dm_{SI}^{-2} - 3a]$$

Thus we find the form of the potential of eqs. 2.1 and 2.3 plus Yukawa-like terms whose small mass m_{SI} – perhaps near zero. As a result the effective potential is

$$V(r) = -2g^2/r + g^2\lambda^2 r \qquad (A\text{-}2.3)$$

as seen in appendix A and in the Charmonium calculations with[41]

$$g^2\lambda^2 = -(2a)^{-2}d/(8\pi) \qquad (5.31)$$
$$-2g^2 = -(2a)^{-2}[3a - 4dm_{SI}^{-2}]/(4\pi) \qquad (5.32)$$

where $g = \sqrt{(\kappa/2)} = 0.552$ and $\lambda = 0.761$ GeV (the result of Charmonium analysis). Thus

---

[41] We note the constant, a, that appears in chapter 4 and this chapter is NOT the Charmonium constant, a, in eq. 2.1.

$$\lambda^2 = d/[4dm_{SI}^{-2} - 3a] \qquad (5.33)$$

We will discuss the relations of the constants within the unified theory in chapter 10.

6. Gravity Sector in our Higher Derivative Theory – MoND!

In the previous chapter we determined the dynamics of the Strong Interaction sector of this unified theory. This theory is a subsector of our Theory of Everything. In this chapter we will develop the gravitation sector of this subsector.[42] We will see that the theory has higher derivative dynamic equations but unlike the Strong Interaction sector, which yields color (quark-gluon) confinement, the gravitation dynamic equations do not have confinement – but do yield a modified form of gravity at intermediate distances of the order of the average galactic radius.

The modification of gravity implied by our theory is consistent with the need for Dark Matter described in our Theory of Everything (Blaha (2015a), (2016c) and (2016g)).[43] It is also consistent with a MoND theory[44] with the addition of sixth order derivatives in the Gravitation sector lagrangian in a manner consistent with the higher order terms appearing in the Strong and Electromagnetic Interactions sectors of the unified theory.[45] *Thus our approach to MoND does not require a major change in Mechanics, quantum theory, or General Relativity (modulo higher order derivatives). And the Euler-Lagrange canonical procedure still yields the dynamical field equations.*

The consistency between the need for higher order derivatives in both the Strong Interaction and Gravitation sectors to obtain agreement with experiment is encouraging.

The development in this chapter is a generalization with higher order derivative terms of the unified theory presented in chapter 6 of Blaha (2016d).

---

[42] There are other higher derivative theories – some with two metrics, and some with a metric plus vector plus scalar field formulation. The present work is based on a unification of Electromagnetic, Weak, Strong and Gravity sectors, and a totally different formalism. Some significant references are: M. Milgrom, Phys. Rev. **D80**, 123536 (2009); C. Skordis et al, Phys. Rev. Lett. **96**, 011301 (2006); R. H. Sanders, Astrophysical Journal **480**, 492 (1997); and references therein; J. D. Bekenstein, Phys. Rev. **D70**, 083509 (2004) and references therein; J-P. Bruneton, Phys. Rev. **D76**, 124012 (2007) uses higher derivative gravity and metric, vector, scalar fields. See also references within these articles.

[43] I. Ferreras et al, Phys. Rev. Lett., **96**, 011301 (2006) shows the need for both Dark Matter and MoND based on studies of astrophysical data.

[44] Our gravity theory has aspects that are virtually identical to the MoND theories described in A. Balakin et al, Phys. Rev. **D70**, 064027 (2004); H-S Zhao et al, Phys. Rev. **D82**, 103001 (2010); and references therein. However our approach is very different.

[45] Chapter 13 describes the presence of a new U(4) General Relativistic gauge field that modifies gravitation at galactic scale distances. This significant force, if it exists, also contributes MoND-like changes to the force of gravity.

6.1 Gravitation Sector Dynamic Equations

Our gravity sector has two metric fields, $g_{\mu\nu}$ and $g^2{}_{\mu\nu}$ derived from the unified formalism described in chapter 4. Some of the relevant gravitation equations found in chapter 4 are:

$$H^\sigma{}_{\nu\mu} = \Gamma^\sigma{}_{\nu\mu} + \Gamma^{2\sigma}{}_{\nu\mu} \quad (4.43a)$$

$$H^\beta{}_{\sigma\nu\mu} = \partial_\mu H^\beta{}_{\sigma\nu} - \partial_\nu H^\beta{}_{\sigma\mu} + H^\gamma{}_{\nu\sigma} H^\beta{}_{\gamma\mu} - H^\gamma{}_{\mu\sigma} H^\beta{}_{\gamma\nu} \quad (4.50a)$$

$$H_{\sigma\mu} = H^\beta{}_{\sigma\beta\mu} \quad (4.59a)$$

$$H = g^{\sigma\mu} H_{\sigma\mu} \quad (4.60a)$$

$$\mathcal{H} = R'^1 + R'^2$$

We use the gravitational sector lagrangian:[46]

$$\mathcal{L}_G = \mathrm{Tr}\,\sqrt{g}[MD_\nu R'^1{}_{G\sigma\mu} D^\nu R'^2{}_G{}^{\sigma\mu} + aR'^1{}_{G\sigma\mu} R'^2{}_G{}^{\sigma\mu} + bH + cg^{\sigma\mu} g^2{}_{\sigma\mu} + c' g^{2\sigma\mu} g^2{}_{\sigma\mu}] \quad (4.65a)$$

where

$$\begin{aligned} D_\nu V_\mu &= (\partial_\nu + iF_\nu)V_\mu - H^\sigma{}_{\nu\mu} V_\sigma \\ &= [g^\sigma{}_\mu \partial_\nu + ig^\sigma{}_\mu F_\nu - H^\sigma{}_{\nu\mu}]V_\sigma \\ &= [g^\sigma{}_\mu \partial_\nu + iD^\sigma{}_{\mu\nu}]V_\sigma \end{aligned} \quad (4.42)$$

where M, a, b, c, and d are constants with[47]

$$a = 1/(2f) = 0.906 \quad (5.11)$$

and

$$b = (16\pi G)^{-1} \quad (4.73)$$

G is Newton's gravitational constant.

The lagrangian dynamic equations following from eq. 4.65a are difficult. We consequently will examine the weak gravitation limiting case where we can approximate the two metrics with

$$g_{\mu\nu} \cong \eta_{\mu\nu} + h_{\mu\nu} \quad (6.1)$$

---

[46] We note the constant, a, that appears in chapters 4, 5, and this chapter is NOT the Charmonium constant, a, in eq. 2.1.

[47] This value of a is obtained in the Charmonium calculation of eq. 2.1.

$$g^2_{\mu\nu} \cong \eta_{\mu\nu} + h^2_{\mu\nu} \tag{6.2}$$

where

$$|h_{\mu\nu}| << 1 \qquad |h^2_{\mu\nu}| << 1 \tag{6.3}$$

Using the relations

$$\partial_\mu h^\mu{}_\nu = ½\partial_\nu h^\mu{}_\mu \tag{6.4}$$
$$\partial_\mu h^{2\mu}{}_\nu = ½\partial_\nu h^{2\mu}{}_\mu \tag{6.5}$$

and neglecting higher order terms in $h_{\mu\nu}$ and $h^2_{\mu\nu}$ we find

$$R'^1_{\mu\nu} = ½[\Box h_{\mu\nu} - \partial_\mu\partial_\lambda h^\lambda{}_\nu - \partial_\nu\partial_\lambda h^\lambda{}_\mu + \partial_\nu\partial_\mu h^\lambda{}_\lambda]$$
$$\cong ½\Box h_{\mu\nu} \tag{6.6}$$
$$R'^1 \cong ½\Box h_\mu{}^\mu$$

$$R'^2_{\mu\nu} = ½[\Box h^2_{\mu\nu} - \partial_\mu\partial_\lambda h^{2\lambda}{}_\nu - \partial_\nu\partial_\lambda h^{2\lambda}{}_\mu + \partial_\nu\partial_\mu h^{2\lambda}{}_\lambda]$$
$$\cong ½\Box h^2_{\mu\nu} \tag{6.7}$$
$$R'^2 \cong ½\Box h^2{}_\mu{}^\mu$$

with

$$D_\nu = \partial_\nu$$

upon neglecting higher order terms.

Substituting in eq. 4.65a above we find the *effective quadratic* part of the lagrangian (in $h_{\mu\nu}$ and $h^2_{\mu\nu}$) is

$$\mathcal{L}_G = \sqrt{g}[M\partial_\nu R'^1{}_{G\sigma\mu}\partial^\nu R'^2{}_G{}^{\sigma\mu} + a\Box h_{\sigma\mu}\Box h^{2\sigma\mu}/4 + ½b(\partial_\alpha h^{\sigma\mu}\partial^\alpha h_{\sigma\mu} + \partial_\alpha h^{2\sigma\mu}\partial^\alpha h^2{}_{\sigma\mu}) + c(4 + \eta^{\sigma\mu}h^2{}_{\sigma\mu} + $$
$$+ h^{\sigma\mu}\eta_{\sigma\mu} + h^{\sigma\mu}h^2{}_{\sigma\mu}) + c'(2\eta^{\sigma\mu}h^2{}_{\sigma\mu} + h^{2\sigma\mu}h^2{}_{\sigma\mu}) + 1/4(h_{\mu\nu} + h^2{}_{\mu\nu})T^{\mu\nu}] \tag{6.8}$$

Using partial integrations, we find the standard Euler-Lagrange canonical technique for determining the equations of motion from a lagrangian for independent variations with respect to $h_{\mu\nu}$ and $h^2_{\mu\nu}$ yields

$$-M\Box^3h^{2\mu\nu}/4 + a\Box^2h^{2\mu\nu}/4 + ½b\Box h^{\mu\nu} + c(\eta^{\mu\nu} + h^{2\mu\nu}) + 1/4T^{\mu\nu} = 0 \tag{6.9}$$

$$-M\Box^3h^{\mu\nu}/4 + a\Box^2h^{\mu\nu}/4 + ½b\Box h^{2\mu\nu} + c(\eta^{\mu\nu} + h^{\mu\nu}) + 2\,c'(\eta^{\mu\nu} + h^{2\mu\nu}) + 1/4T^{\mu\nu} = 0 \tag{6.10}$$

The term $c\eta^{\mu\nu}$ can be viewed as part of the total energy-momentum tensor $T'^{\mu\nu}$:

$$T'^{\mu\nu} = T^{\mu\nu} + 2c\eta^{\mu\nu} \tag{6.10a}$$

It plays a role similar to the Cosmological Constant. Subtracting the equations we find

$$M(\Box^3 h^{\mu\nu}/4 - \Box^3 h^{2\mu\nu}/4) + a\Box^2 h^{2\mu\nu}/4 - a\Box^2 h^{\mu\nu}/4 + ½b\Box h^{\mu\nu} - ½b\Box h^{2\mu\nu} + c(h^{2\mu\nu} - h^{\mu\nu}) - 2\,c'(\eta^{\mu\nu} + h^{2\mu\nu}) = 0 \quad (6.11)$$

and thus

$$[-M\Box^3 + a\Box^2 - ½b\Box + (4c - 8\,c')]h^{2\mu\nu} = 8\,c'\eta^{\mu\nu} - M\Box^3 h^{\mu\nu} + a\Box^2 h^{\mu\nu} - ½b\Box h^{\mu\nu} + 4ch^{\mu\nu} \quad (6.12)$$

Therefore we determine the metric equation for $h^{2\mu\nu}$

$$h^{2\mu\nu} = [-M\Box^3 + a\Box^2 - ½b\Box + (4c - 8\,c')]^{-1}[8e\eta^{\mu\nu} - M\Box^3 h^{\mu\nu} + a\Box^2 h^{\mu\nu} - ½b\Box h^{\mu\nu} + 4ch^{\mu\nu}] \quad (6.13)$$

Substituting in eq. 6.9 we obtain

$$[-M\Box^3/4 + a\Box^2/4 + c][-M\Box^3 + a\Box^2 - ½b\Box + (4c - 8\,c')]^{-1}[8e\eta^{\mu\nu} - M\Box^3 + a\Box^2 h^{\mu\nu} - ½b\Box h^{\mu\nu} + 4ch^{\mu\nu}] + + ½b\Box h^{\mu\nu} + 1/4T'^{\mu\nu} = 0 \quad (6.14)$$

Eq. 6.14 becomes with simple changes:

$$\{[-M\Box^3/4 + a\Box^2/4 + c][-M\Box^3 + a\Box^2 - ½b\Box + (4c - 8\,c')]^{-1}[-M\Box^3 + a\Box^2 - ½b\Box + 4c] + + ½b\Box\}h^{\mu\nu} = -1/4T'^{\mu\nu} \quad (6.15)$$

6.2 Gravity Potential

Assuming that we are dealing with non-relativistic matter we can calculate the gravity potential contribution from eq. 6.15

$$V_{G1}(\mathbf{x}) = -\int d^3k\, \exp(i\mathbf{k}\cdot\mathbf{x})V_{G1}(\mathbf{k})/(2\pi)^3 \quad (6.16)$$

where

$$V_{G1}(\mathbf{k}) = \{[M\mathbf{k}^6/4 + a\mathbf{k}^4/4 + c][M\mathbf{k}^6 + a\mathbf{k}^4 + ½b\mathbf{k}^2 + (4c - 8\,c')]^{-1}[M\mathbf{k}^6 + a\mathbf{k}^4 + ½b\mathbf{k}^2 + 4c] - - ½b\mathbf{k}^2\}^{-1}$$

$$= [M\mathbf{k}^6 + a\mathbf{k}^4 + ½b\mathbf{k}^2 + (4c - 8c')]/\{[M\mathbf{k}^6/4 + a\mathbf{k}^4/4 - ½b\mathbf{k}^2 + c][M\mathbf{k}^6 + a\mathbf{k}^4 + ½b\mathbf{k}^2 + 4c] + + 4\,c'b\mathbf{k}^2\} \quad (6.17)$$

Similarly eq. 6.13 implies the other contribution to the total gravity potential is

$$V_{G2}(\mathbf{x}) = -\int d^3k\, \exp(i\mathbf{k}\cdot\mathbf{x})V_{G2}(\mathbf{k})/(2\pi)^3 \quad (6.18)$$

where

$$V_{G2}(\mathbf{k}) = \{[Mk^6 + ak^4 + ½bk^2 + 4c + 8\,c']^{-1}[Mk^6 + a\,k^4 + ½bk^2 + 4c - 8\,c']\}V_{G1}(\mathbf{k}) \quad (6.19)$$

The total gravity potential energy is thus

$$V^{tot}{}_G(\mathbf{x}) = V_{G1}(\mathbf{x}) + V_{G2}(\mathbf{x}) \quad (6.20)$$

6.3 Solution for the Case of a Massless Graviton

Eqs. 6.16-6.17 can generate a massless graviton, which seems a requirement based on cosmological considerations, and also generate a pair of massive gravitons of very low mass. The massive gravitons generate a MoND-like potential that might explicate the anomalous gravitation effects seen at distances of the order of galactic dimensions.

If we set the 'Cosmological Constants' c = c' = 0, then eq. 6.17 becomes

$$V_{G1}(\mathbf{k}) = 4/\{M\mathbf{k}^2[\mathbf{k}^4 + a\mathbf{k}^2/M - 2b/M]\} \quad (6.21)$$

The denominator of eq. 6.21 can be factored into the form

$$V_{G1}(\mathbf{k}) = 4/\{M\mathbf{k}^2[\mathbf{k}^2 + ½m_{SI}^2 + ½(m_{SI}^4 + 8bm_{SI}^2/a)^{½}][\mathbf{k}^2 + ½m_{SI}^2 - ½(m_{SI}^4 + 8bm_{SI}^2/a)^{½}]\} \quad (6.22)$$

where

$$m_{SI} = (a/M)^{½} \quad (5.20)$$

Assuming, as shown in chapter 10, $m_{SI}^2 << 8b/a$, or $a^2/8 < 1/8 << bM = M(16\pi G)^{-1}$, which is reasonable since a is approximately one, we see

$$V_{G1}(\mathbf{k}) \cong 4/\{M\mathbf{k}^2[\mathbf{k}^2 + (2b/a)^{½}][\mathbf{k}^2 - (2b/a)^{½}]\} \quad (6.23)$$
$$= [2/(ba)]\{-2/\mathbf{k}^2 - 1/[\mathbf{k}^2 + (2b/a)] + 1/[\mathbf{k}^2 - (2b/a)]\}$$

up to negligible terms.

From eq. 6.19 we see $V_{G2}(\mathbf{k}) = V_{G1}(\mathbf{k})$ if c = c' = 0. Thus the total gravitational potential by eq. 6.8 (in momentum space) is

$$V^{tot}{}_{Gt}(\mathbf{k}) = ½(V_{G1}(\mathbf{k}) + V_{G2}(\mathbf{k})) \quad (6.24)$$
$$= [2/(ba)]\{-2/\mathbf{k}^2 - 1/[\mathbf{k}^2 + (2b/a)] + 1/[\mathbf{k}^2 - (2b/a)]\}$$
$$\cong 8\pi G\{-2/\mathbf{k}^2 - 1/[\mathbf{k}^2 + (2b/a)] + 1/[\mathbf{k}^2 - (2b/a)]\}$$

and the coordinate space potential to leading order is[48]

$$V^{tot}{}_G(\mathbf{r}) = -G/r - a_1Ge^{-m_Gr}/r + a_2G\cos(m_Gr)/r \quad (6.25)$$

[48] Since the theory has higher order drivatives that could lead to unitarity problems Feynman propagators must be taken in Principal order. Since potentials are a part of Feynman propagators the potentials' real value must be used.

where

$$m_G^2 = 2b/a \cong 10^{55}\ ev^2 = 10^{27}\ GeV^2 \qquad (6.25a)$$
$$a_1 = ½$$
$$a_2 = ½$$

6.3.1 Small and Ultra-Large behavior of the Gravitational Potential

In chapter 10 we will see that m_G is an extremely small mass. For small distances

$$V^{tot}{}_G(\mathbf{r}) \cong -G/r \qquad (6.25b)$$

For distances very much larger than tens of thousands of light years (inter-galactic distances)

$$V^{tot}{}_G(\mathbf{r}) \cong -G/r + ½G\cos(m_G r)/r \qquad (6.25c)$$

Eq. 6.25c shows an oscillatory behavior around –G/r for distances of the order of billions of light years that would suggest 'clumping' of galaxies in some regions and voids between the clumps. This feature seems to have been observed in the cosmos in galactic surveys.

6.3.2 MoND Behavior of Gravitation at Distances of the Order of Galactic Radii

In this subsection we discuss the gravity potential at distances of the order of tens of thousands of light years. Note that the approximate real part of the expansion of the third term in eq. 6.25 to third order yields

$$V^{tot}{}_G(\mathbf{r}) \sim -G/r + a_1 G m_G{}^3 r^2 - a_2 G m_G{}^2 r + \text{constants} \qquad (6.26)$$

with the resultant force

$$\mathbf{F} = \nabla V^{tot}{}_G(\mathbf{r}) \sim G\mathbf{r}/r^3 + 2a_1 G m_G{}^3 \mathbf{r} - a_2 G m_G{}^2 \mathbf{r}/r + \ldots \qquad (6.27)$$

This result is to be compared to the MoND force of A. Balakin et al, Phys. Rev. **D70**, 064027 (2004):[49]

$$F = -\lambda G m[M/r^2 - |\Pi_c|\, r/c^2] \qquad (6.28)$$

and the vector form suggested by H-S Zhao et al, Phys. Rev. **D82**, 103001 (2010):

$$\partial\Phi/\partial\mathbf{r} = Gm\mathbf{r}/r^3 + (Gm)^{½}\mathbf{r}/r^2 \qquad (6.29)$$

[49] The constant c in eq. 6.28 is the speed of light, and M is the mass (not the M used in our lagrangian equations.)

The resemblance to MoND estimates is clearly striking in both of the above comparison cases.

Recent studies of 153 galaxies confirm the MoND discrepancy from Newtonian gravitation.[50]

We will see in chapter 10 that we can obtain values for M using MoND-like distances ranging from the order of 20,000 to 100,000 light years (the radius of the Andromeda galaxy). These values of M will cause the gravity potential at these distance ranges to be increased by Yukawa like terms with small values of m_G.[51] In addition the masses in the Yukawa-like parts of the Strong Interaction potential will be extremely small but their effects will be 'shielded' so eqs. 2.1 and 2.3 will yield the 'known' effective Strong Interaction potential.

Thus our unified theory resolves the MoND galaxy gravity problem with a correct choice of M and maintains the Strong Interaction potential found for Charmonium. A modification of Newton's law then becomes unnecessary (and udesirable in the author's view.)

---

[50] S. S. McGaugh, F. Lelli, and J. M. Schombert, arXiv: 1609.0591 (2016).

[51] The third term in eq. 6.25 is an oscillating Yukawa term that, because of the smallness of m_G, is slowly varying towards the end of a galaxy and thus could be well within observational error bounds. It appears that the real part of the third term is the contribution to the total gravitational potential using Principal value propagators.

7. Electromagnetic Sector of GEMS

7.1 Electromagnetic Sector Lagrangian and Free Field Propagators

The electromagnetic lagrangian density term is

$$\mathcal{L}_E = \sqrt{g}\{aF^1{}_{E\sigma\mu}F^2{}_E{}^{\sigma\mu}\} \tag{4.68a}$$

The quadratic part of $\mathcal{L}_E$ from which the free field Feynman propagators are obtained is:

$$\mathcal{L}_{Efree} = aF^1{}_{E\sigma\mu}F^2{}_E{}^{\sigma\mu} \tag{7.1}$$

with the metric determinant, g, set to 1 where

$$F^1{}_{E\mu\kappa} = \partial A_E{}^1{}_\mu/\partial x^\kappa - \partial A_E{}^1{}_\kappa/\partial x^\mu \tag{7.2}$$
$$F^2{}_{E\mu\kappa} = \partial A_E{}^2{}_\mu/\partial x^\kappa - \partial A_E{}^2{}_\kappa/\partial x^\mu$$

The conjugate momenta to $A^1{}_\mu$ and $A^2{}_\mu$ are respectively

$$\pi^1{}_\mu = \partial\mathcal{L}_E/(\partial A_E{}^1{}_\mu/\partial t) = aF_E{}^{2\mu t} \tag{7.3}$$
$$\pi^2{}_\mu = \partial\mathcal{L}_E/(\partial A_E{}^2{}_\mu/\partial t) = aF_E{}^{1\mu t}$$

The non-zero, equal time commutation relations are

$$[\pi^i{}_\mu(\mathbf{x}, t), A_E{}^j{}_\nu(\mathbf{y}, t)] = i(1 - \delta^{ij})\delta^{G(\mu\nu)}(\mathbf{x} - \mathbf{y}) \tag{7.4}$$

where i and j label the fields, and G(μν) indicates the gauge[52] G and the associated index expressions, with

$$\delta^{G(\mu\nu)}(\mathbf{x} - \mathbf{y}) = \int d^4k \exp(-ik\cdot x)b^{G\mu\nu}(k)/(2\pi)^4 \tag{7.5}$$

where $b^{G\mu\nu}(k)$ is a polynomial in k with a δ-function factor restricting the integration over k.

7.2 Dynamical Equations

The Euler-Lagrange dynamical equations (in the Landau gauge) due to independent variations with respect to $A_E{}^1{}_\mu$ is

[52] Not the gravitational coupling constant.

$$2a\Box A_E{}^2{}_\mu = -e\bar{\psi}\gamma_\mu\psi \tag{7.6}$$

and with respect to $A_E{}^2{}_\mu$ is

$$2a\Box A_E{}^1{}_\mu = -e\bar{\psi}\gamma_\mu\psi \tag{7.7}$$

Subtracting the equations we find

$$-2a\Box A_E{}^1{}_\mu + 2a\Box A_E{}^2{}_\mu = 0 \tag{7.8}$$

or

$$A_E{}^2{}_\mu = [2a\Box]^{-1}[2a\Box A_E{}^1{}_\mu] = A_E{}^1{}_\mu \tag{7.9}$$

with the result

$$-2a\Box A_E{}^1{}_\mu = -e\bar{\psi}\gamma_\mu\psi \tag{7.10}$$

Eq. 7.10 leads to the Principal Value (Feynman) propagator:

$$\begin{aligned} D^{12}{}_{\mu\nu}(x-y) &= P\, -i\langle 0|T(A_E{}^1{}_\mu(x),\, A_E{}^1{}_\nu(y)|0\rangle \\ &= P\int d^4k\, \exp(-ik\cdot x)b_{\mu\nu}(k)D(k)/(2\pi)^4 \\ &= D^{21}{}_{\mu\nu}(x-y) \end{aligned} \tag{7.11}$$

where $b_{\mu\nu}(k)$ is a Landau gauge polynomial in k, and

$$D(k) = \{-2ak^2\}^{-1} = -(1/2a)[1/k^2] \tag{7.12}$$

Due to the form of the electromagnetic interaction the total effective interaction is

$$\begin{aligned} D^{tot}{}_{\mu\nu}(x-y) &= D^{21}{}_{\mu\nu}(x-y) + 2D^{12}{}_{\mu\nu}(x-y) \\ &= P\int d^4k\, \exp(-ik\cdot x)b_{\mu\nu}(k)\{-(1/a)[1/k^2]\} \end{aligned} \tag{7.13}$$

Since a = 0.90 by eq. 5.11 in section 6.1 from the Charmonium spectrum fit of the Cornell group and, since a value of a = 1 is not precluded in view of the problems that have recently surfaced in the higher states of charmonium, we choose to believe a = 1. Thus we believe

$$D^{tot}{}_{\mu\nu}(x-y) = P\int d^4k\, \exp(-ik\cdot x)b_{\mu\nu}(k)\{-1/k^2\} \tag{7.14}$$

7.3 Electromagnetic Interaction Potential

Eq. 7.13 leads to the familiar form of the total Electromagnetic potential (in Principal Value!):

$$V_E(\mathbf{r}) = \alpha/r \tag{7.15}$$

The author notes the common value of a determined by the Cornell group Charmonium calculation $a_{charmonium} = 0.90$ *(which is consistent with one),and the value of a thus obtained in the electromagnetic sector that yields the known Coulomb potential, and the value of a in the Storng interaction sector. Tthis consistency argues strongly for the unification of the four interactions presented here.*

8. Weak Interaction Sector

The Weak interaction Yang-Mills vector boson lagrangian terms with the metric determinant g = 1 are

$$\mathcal{L}_{SU(2)} = \text{Tr}\ aF_W{}^1{}_{\sigma\mu}F_W{}^{2\sigma\mu} \tag{8.1}$$

by eqs. 4.81 and 4.84 with

$$a = -\tfrac{1}{2} \tag{8.2}$$

where

$$F_W{}^1{}_{\kappa\mu} = \partial W^1{}_\mu/\partial x^\kappa - \partial W^1{}_\kappa/\partial x^\mu + ig_W[W^1{}_\kappa, W^1{}_\mu] \tag{8.3}$$

$$F_W{}^2{}_{\kappa\mu} = \partial W^2{}_\mu/\partial x^\kappa - \partial W^2{}_\kappa/\partial x^\mu + ig_W[W^2{}_\kappa, W^2{}_\mu] + ig_W[W^1{}_\kappa, W^2{}_\mu] + ig_W[W^2{}_\kappa, W^1{}_\mu]$$

The derivation of the dynamical equations after the introduction of fermion terms interacting with the Weak vetor bosons is straightforward (See, for example, Blaha(2015a).) and will not be repeated here.

9. The Interaction Lagrangian

The interaction terms of the unified four interaction lagrangian are given by eq. 4.82 if we neglect the interactions with gravitation due to the overall metric determinant factor $\sqrt{g}$. In this chapter we will describe the interactions amongst the four interactions excepting those simply due to the $\sqrt{g}$ factor. The 'full' interactions amongst interactions lagrangian terms $\mathscr{L}_{int}$ are then:[53,54]

$$\begin{aligned}\mathscr{L}_{int} = \mathrm{Tr}\,\sqrt{g}[M\{&-i({A_E}^1{}_\nu + {A_E}^2{}_\nu + W^1{}_\nu + W^2{}_\nu)F^1{}_{\sigma\mu}D^\nu F^{2\sigma\mu} - iD_\nu F^1{}_{\sigma\mu}({A_E}^{1\nu} + {A_E}^{2\nu} + \\ &+ W^{1\nu} + W^{2\nu})\, F^{2\sigma\mu} + iD_\nu R'^1{}_{G\sigma\mu}[\partial^\nu + i(A^{1\nu} + A^{2\nu} + {A_E}^{1\nu} + {A_E}^{2\nu} + W^{1\nu} + W^{2\nu})]F^{2\sigma\mu} + \\ &+ i[\partial_\nu + i(A^1{}_\nu + A^2{}_\nu + {A_E}^1{}_\nu + {A_E}^2{}_\nu + + W^1{}_\nu + W^2{}_\nu)]F^1{}_{\sigma\mu}D^\nu R'^2{}_G{}^{\sigma\mu}\} + \\ + a\{&-{F_E}^1{}_{\sigma\mu}({F_W}^{2\sigma\mu} + F^{2\sigma\mu}) - F^1{}_{\sigma\mu}({F_E}^{2\sigma\mu} + {F_W}^{2\sigma\mu}) - {F_W}^1{}_{\sigma\mu}({F_E}^{2\sigma\mu} + F^{2\sigma\mu}) + \\ &+ iR'^1{}_{G\sigma\mu}({F_E}^{2\sigma\mu} + {F_W}^{2\sigma\mu} + F^{2\sigma\mu}) + i({F_E}^1{}_{\sigma\mu} + {F_W}^1{}_{\sigma\mu} + F^1{}_{\sigma\mu})R'^2{}_G{}^{\sigma\mu}\}]\end{aligned} \tag{4.82}$$

where D_ν is given by eq. 4.51.

$\mathscr{L}_{int}$ has electromagnetic-Strong interaction terms, gravitation-Strong interaction terms, and gravitation-electromagnetic terms, Weak-Strong terms, Weak-electromagnetic terms, and Weak-Gravitation terms.

$$\mathscr{L}_{int} = \mathscr{L}_{intEM\text{-}S} + \mathscr{L}_{intE\text{-}G} + \mathscr{L}_{intS\text{-}G} + \mathscr{L}_{intS\text{-}W} + \mathscr{L}_{intG\text{-}W} + \mathscr{L}_{intE\text{-}W} \tag{9.1}$$

Since $\mathscr{L}_{int}$ is a trace it simplifies the interactions significantly.

9.1 Electromagnetic–Strong Interaction

From eq. 4.82 we see that there are important terms with Electromagnetic and Strong interaction field factors. After taking the trace we find

$$\mathscr{L}_{intEM\text{-}S} = -\mathrm{Tr}\; iM\{({A_E}^1{}_\nu + {A_E}^2{}_\nu)\, F^1{}_{\sigma\mu}D^\nu F^{2\sigma\mu} + iD_\nu F^1{}_{\sigma\mu}({A_E}^{1\nu} + {A_E}^{2\nu})F^{2\sigma\mu}\} \tag{9.2}$$

with $g = 1$. $\mathscr{L}_{intEM\text{-}S}$ generates a combined photon-gluon vertex insertion in gluon interactions between quarks within a hadron. Figs. 9.1 and 9.2 show two possible vertex insertions in a gluon line.

---

[53] The coupling constants of the gauge fields are not shown in the interests of simplicity. See eq. 4.83 for the coupling constants, which will be treated as implicit in the gauge fields in this chapter.

[54] It is understood that the traces are taken separately for SU(2) and SU(3) matrices.

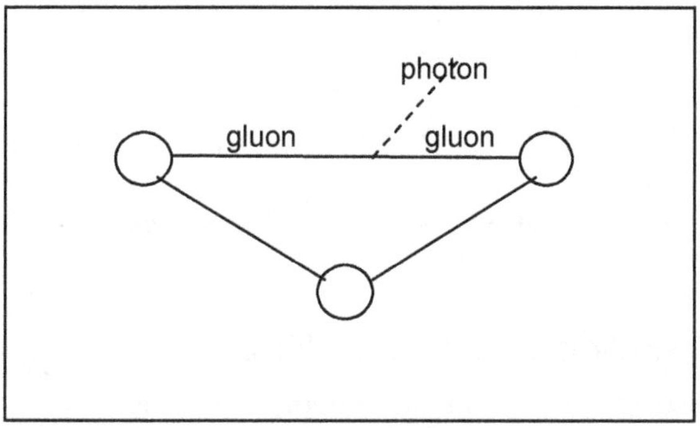

Figure 9.1. A single 'outgoing' photon vertex insertion in a gluon line. Only single gluon lines between the three quarks are displayed.

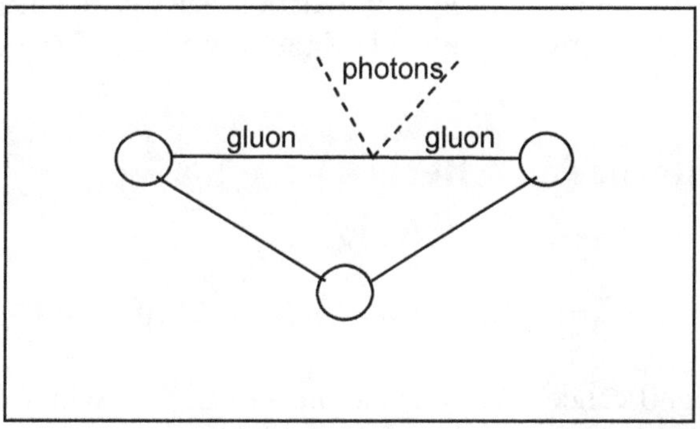

Figure 9.2. An 'outgoing' two photon vertex insertion in a gluon line. Only single gluon lines between the three quarks are displayed.

The gluon line, by itself, has $1/k^4$ and $1/k^2$ momentum space propagator terms. *The insertion of the vertex in Fig. 9.1 generated by* $\mathcal{L}_{intEM-S}$ *yields a combined momentum factor of* $k^3(k^4k^4)^{-1} = k^{-5}$ *which would make it (summed over all gluon lines) a significant contribution to the proton spin determination in deep inelastic electron-nucleon scattering.*[55] *The insertion of the vertex in Fig. 9.2 generated by* $\mathcal{L}_{intEM-S}$ *yields a combined momentum factor of* $k^2(k^4k^4)^{-1} = k^{-6}$ *which may have a less significant effect.*

[55] See C. A. Aidala, S. D. Bass, D. Hasch, and G. K. Mallot, arXiv: 1209.2803v2 (2013) and references therein for a review of the 'missing' nucleon spin puzzle.

9.2 Nucleon Spin Puzzle

Thus our unified theory may also account solve the nucleon spin puzzle as well as quark confinement and MoND deviations in gravity at the galactic scale.The interactions in Figs. 9.1 and 9.2 introduce a direct connection between photons and spin one gluons. Thus their contributions to the summations of proton spin interactions may account for the 'missing' two-thirds of proton spin.

Our unified theory has a gluon-photon interaction that is not found in the conventional Standard Model.

9.3 Electromagnetic−Gravitation Interaction Effects

The terms that generate electromagnetic-graviton interaction vertices are

$$\mathcal{L}_{intE\text{-}G} = a\sqrt{g}\{R'^{1}{}_{G\sigma\mu}F^{2}{}_{E}{}^{\sigma\mu} + F^{1}{}_{E\sigma\mu}R'^{2}{}_{G}{}^{\sigma\mu}\} \qquad (9.3)$$

by eq. 4.82. These interaction terms may play a role in the dynamics and equation of state of stars with significant electromagnetic fields, and large gravitation, such as neutron and quark stars. Divergences generated via these terms can be made finite using Two-Tier quantum field theory as described in our earlier books.

9.4 Gravitation−Strong Interaction Effects

Graviton-gluon interaction vertices are generated by:

$$\mathcal{L}_{intG\text{-}S} = \text{Tr}\, M\sqrt{g}\{D_\nu R'^{1}{}_{G\sigma\mu}[\partial^\nu + i(A^{1\nu} + A^{2\nu})]F^{2\sigma\mu} + [\partial_\nu + i(A^{1}{}_{\nu} + A^{2}{}_{\nu})]F^{1}{}_{\sigma\mu}D^\nu R'^{2}{}_{G}{}^{\sigma\mu}\} \qquad (9.4)$$

These terms are not normally of significance since gravitation is negligible within hadrons, and the Strong interaction, being confined, is negligible outside of hadrons. In addition taking the trace in eq. 9.4 will eliminate some of the terms.

However these interaction terms may play a role in the dynamics and equation of state of quark stars.[56] Divergences generated via these terms can be made finite using Two-Tier quantum field theory as described in our earlier books.

9.5 Gravitation−Electromagnetic−Strong Interaction Vertex Effects

There are no interaction vertices with this combination of three interaction field factors after the trace is taken.

$$\mathcal{L}_{intG\text{-}E\text{-}S} = \text{Tr}\,\sqrt{g}M\{-D_\nu R'^{1}{}_{G\sigma\mu}(A_E{}^{1\nu} + A_E{}^{2\nu})F^{2\sigma\mu} - (A_E{}^{1}{}_{\nu} + A_E{}^{2}{}_{\nu})F^{1}{}_{\sigma\mu}D^\nu R'^{2}{}_{G}{}^{\sigma\mu}\} = 0 \qquad (9.5)$$

---

[56] Quark-gluon plasmas do not have significant gravitational effects within them. Quark stars may have both significant Strong and gravitational fields. Thus these interactions may be of significance within quark stars.

Note: Taking the trace in eq. 9.5 will eliminate some of the terms.

9.6 Weak Interaction–Strong Interaction Effects

The terms in eq. 4.82 that generate the interplay of these interactions are

$$\begin{aligned}\mathcal{L}_{\text{intS-W}} = \text{Tr}\,\sqrt{g}[M\{&-i(A^1{}_v + A^2{}_v)F^1{}_{\sigma\mu}i(W^{1v} + W^{2v})F^{2\sigma\mu} - \\ &- i(W^1{}_v + W^2{}_v)F^1{}_{\sigma\mu}[\partial^v + i(W^{1v} + A^{1v} + W^{2v} + A^{2v})]F^{2\sigma\mu} - \\ &- i[\partial_v + i(A^1{}_v + A^2{}_v + W^1{}_v + W^2{}_v)]F^1{}_{\sigma\mu}(W^{1v} + W^{2v})F^{2\sigma\mu} + \\ &+ (W^1{}_v + W^2{}_v))F^1{}_{\sigma\mu}(A^{1v} + A^{2v})F^{2\sigma\mu}\} + a\{-F^1{}_{\sigma\mu}F_W{}^{2\sigma\mu} - F_W{}^1{}_{\sigma\mu}F^{2\sigma\mu}\}]\end{aligned} \qquad (9.6)$$

Some of these terms disappear when the trace is taken. These terms, combined with fermion terms, may be a source of anomalous CP violation as well as other violations.[57] They do not appear in The Standard Model as commonly depicted.

9.7 Weak–Gravitation Interaction Effects

The interaction terms with solely Weak and Gravitation terms are

$$\begin{aligned}\mathcal{L}_{\text{intG-W}} = \text{Tr}\,\sqrt{g}[M\{&-D_vR'^1{}_{G\sigma\mu}(W^{1v} + W^{2v})F^{2\sigma\mu} - (W^1{}_v + W^2{}_v)F^1{}_{\sigma\mu}D^vR'^2{}_G{}^{\sigma\mu}\} + \\ &+ ia\{R'^1{}_{G\sigma\mu}F_W{}^{2\sigma\mu} + F_W{}^1{}_{\sigma\mu}R'^2{}_G{}^{\sigma\mu}\}]\end{aligned} \qquad (9.7)$$

These terms could play a role in neutrino emission in the vicinity of Black holes or ultra-large stars. Taking the trace in eq. 9.7 will eliminate some of the terms.

9.8 Weak–Electromagnetic Interaction Effects

These interaction terms are zero upon taking the trace.

$$\mathcal{L}_{\text{int}} = \text{Tr}\, a\sqrt{g}\{-F_E{}^1{}_{\sigma\mu}F_W{}^{2\sigma\mu} - F_W{}^1{}_{\sigma\mu}F_E{}^{2\sigma\mu}\} = 0 \qquad (9.8)$$

9.9 Four Interaction Vertex?

Examination of eq. 4.82 shows that no terms survive after taking the trace with respect to SU(2) and SU(3) matrices.

---

[57] Taking the trace in eq. 9.6 will eliminate some of the terms.

10. The Parameters of the Complete Unified Gravitation, Elelectromagnetic, Weak, and Strong Interaction Theory

10.1 Strong Interaction Constants Suggested by Charmonium Studies

The "Cornell group" developed an apparently satisfactory[58] charmed quark bound state spectrum in 1974-5 using a combination of a linear and a 1/r potential as the strong interaction. In a recent fit[59] they gave the potential energy: electrogravistrong theory

$$V(r) = -\kappa/r \; + r/a_{Cornell}{}^{2} \qquad (10.1)$$

where $\kappa = 0.61$, $a_{Cornell} = 2.38\ \text{GeV}^{-1}$ and the charmed quark mass was 1.84 GeV. Substituting these values in eqs. 5.26 – 5.28 and A-2.3 gives

$$g \equiv f = \sqrt{(\kappa/2)} = 0.552 \qquad (10.2)$$
$$\lambda = 0.761\ \text{GeV} \qquad (10.3)$$
$$d = \tfrac{1}{2}\lambda^2/f^2 = 0.950\ \text{GeV}^2 \qquad (10.4)$$
$$a = 1/(2f) = 0.906 \qquad (10.5)$$

The Cornell charmonium potential emerges directly from our theory (appendices A and B). We note the closeness of a[60], and d, to unity suggests that they may be one in value.

$$d = \tfrac{1}{2}\lambda^2/f^2 = 1.0 \quad ??? \qquad (10.6)$$
$$a = 1/(2f) = 1.0 \quad ??? \qquad (10.7)$$

and thus

$$g \equiv f = \tfrac{1}{2} \qquad ??? \qquad (10.8)$$

Then the Cornell values would become

$$\kappa = \tfrac{1}{2} \qquad (10.9)$$
$$a_{Cornell} = (f\lambda)^{-1} = 2.828\ \text{GeV}^{-1} \quad ??? \qquad (10.10)$$

which are not very dissimular to the stated values, and might be accomodated by a shift in the mass of the charmed quark. We leave that to a future study.

---

[58] As did a Harvard group.

[59] E. J. Eichten, K. Lane, and C. Quigg, arXiv:hep-ph/ 0206018 (2002). See this paper for references to earlier work by the "Cornell group" and the "Harvard group" as well as papers by other researchers.

[60] A value of a = 1 would make the coefficient of the 'quartic' term $R'^{1}{}_{G\sigma\mu}R'^{2}{}_{G}{}^{\sigma\mu}$ in eq. 4.65 unity.

We note that $f^2/4\pi = 0.024$ is only a factor of 3.3 more than the fine structure constant – approximately 1/137 = 0.0073. *Therefore perturbative corrections to our inter-quark potential may not be significant and our theory may be the correct theory of the strong interaction. The strong interaction potential in this charmonium fit suggests that the unperturbed potential of the theory presented in Appendices A and B may be a good approximation to the exact potential determined in perturbation theory.*

The smallness of the Strong Interaction terms in the Cornell group potential (eq. 2.1) looks puzzling at first glance. Why is it not large ("Strong")? We believe the strength of the Strong Interaction does not originate in the value of the coupling constant but rather in the linear potential term which provides confinement. The Cornell group potential's 'small' coupling constant then becomes understandable within the context of Strong Interaction.

The remaining constant in eq. 4.88 is M. We will determine this constant in section 10.2. Its successful determination, which links the Strong Interactions and Gravitation, will further support our case for a unified theory.

10.2 Gravitation Constants Suggested by Galactic Scale Deviations from Newtonian Gravity

Since we take c = c' = 0 in chapter 4, the remaining constant to be determined, that affects both the Strong Interactions and Gravitation, is M. In this section we will first determine M by requiring it to influence galactic motions at radii of the order of 100,000 light years (the radius of the Andromeda galaxy).[61] Then we will examine other choices for M.

10.2.1 100,000 Light Year Case

Transforming all distances to electron volts (ev) using $1\ \text{ev}^{-1} = 10^{-6}$ m we find 100,000 ly (light years) $\equiv 3\times10^{-27}\ \text{ev}^{-1}$ implies an approximate graviton upper mass estimate of 3×10^{-27} ev if the gravitation potential and force terms are to modify gravity at galactic distances of the order of tens of thousands of light years.

$$V^{tot}{}_G(\mathbf{r}) = -G/r + a_1 G e^{-m_G r}/r + a_2 G\cos(m_G r)/r \qquad (6.25)$$

$$\mathbf{F} = \nabla V^{tot}{}_G(\mathbf{r}) \sim G\mathbf{r}/r^3 + 2a_1 G m_G{}^3\mathbf{r} - a_2 G m_G{}^2\mathbf{r}/r + \ldots \qquad (6.27)$$

with $a_1 = a_2 = ½$.

Setting the non-zero- graviton masses:

$$m_G{}^2 = (2bm_{SI}{}^2/a)^{½} = (2b/M)^{½} = (3\times10^{-27})^2\ \text{ev}^2 = 10^{-55}\ \text{ev}^2 = (3\times10^{-27}\ \text{ev})^2 \qquad (10.11)$$

and using

$$b = (16\pi G)^{-1} \qquad (4.92)$$

[61] All coupling constant values are based on data from K. A. Olive et al (Particle Data Group), Chinese Physics **C38**, 090001 (2014).

we find

$$M \sim 2\times10^{161}\ \text{ev}^{-2} \tag{10.12}$$

and

$$m_{SI} = (a/M)^{1/2} \cong 10^{-80}\ \text{ev} \cong 10^{-71}\ \text{GeV} \tag{5.20}$$
$$m_G = (2b/a)^{1/2} \cong 10^{14}\ \text{GeV}$$

10.2.2 Galaxy Dynamics Linked to Quark Dynamics

Thus the higher order terms in the total lagrangian, together with a need to have MoND behavior lead to some very small mass, Srong interaction propagator terms, and a very large mass for some graviton propagator terms. This combination of features appears physically reasonable.Thus we have a link between Galaxy dyamics and quark dynamics through the use of a generalized Riemann-Christoffel curvature tensor.

10.2.3 Other MoND Scale Choices for the M

For other choices of the galactic MoND scale ranging from 20,000 – 100,000 light years we find changes in the MoND scale do not significantly affect the order of magnitude of M, m_{SI}, and m_G.

The extremely large size of M leads to extremely small values of the Strong Interaction and Gravitation non-zero masses and, potentially, a large gluon contribution to the missing proton spin. In the case of the Strong Interactions the mass may be well approximated by zero since the regions with color are extremely small as well. Thus the Yukawa-like mass terms are effectively $1/k^4$ and $1/k^2$ terms.

In the case of Gravitation, the absence of color confinement enables the Yukawa-like terms to influence the strength of gravity making MoND (Modified Newtonian Dynamics) possible at galactic distances of the order of tens of thousands of light years while reserving the 1/r Newtonian potential at short distances and cosmic distances much beyond galactic distances.

Thus we have achieved a unification of the four interactions that unites quark confinement, and the MoND effect in Gravitation.

10.3 Weak Interaction Coupling Constant

The Weak Interaction coupling constant (eq. 7.30 in Appendix A) is

$$g_W = 0.619 \tag{10.13}$$

It has an order of magnitude similarity to the other coupling constants listed in eqs. 10.2 – 10.8. Thus there is a consistency in coupling constant values that encourages some confidence in a unified theory of the four known interactions.

11. Three 'New' Interactions

11.1 The Need for More Interactions

The four well-established interactions used earlier to develop a unified quantum field theory formalism do not appear to be the only particle interactions. A variety of extensions of the set of interactions have been proposed – usually with a justification based on symmetry considerations.

In this chapter we will consider three additional interactions based partly on known features of elementary particle physics as well as symmetry considerations. These interactions have been proposed by the author in earlier books in the past two years. In this chapter we will describe them with a view to creating a unified theory of the seven interactions in chapter 12.

The basis of these additional interactions is 1) our existence in a fundamentally complex-valued space (described in earlier books) that is constrained by symmetry breaking to the real-valued space-time that we experience, and 2) the existence of additional symmetries following from apparently conserved quantities such as baryon number that lead to the observed fermion generations and layers of sets of fermion generations. Overall, these considerations lead us beyond The Standard Model of SU(3)⊗SU(2)⊗U(1) to an enhanced Standard Model with SU(3)⊗SU(2)⊗U(1)⊗U(4)⊗U(4)⊗U(4) symmetry.[62]

Most of the material in the following sections of this chapter appears in volumes 1 through 9 of the Physics is Logic series, which appeared in 2015 and the first four months of 2016.

11.2 Curved, Complex 4-Dimensional Space Reality Group and its U(4) Interactions

The Reality group of curved, complex Space is a local U(4) group that has associated Yang-Mills gauge fields that yield interactions with other gauge fields and fundamental fermions.[63] This local U(4) Reality group has 16 generators. An element of the group, U(x), can be expressed in a coordinate system with coordinates x as

---

[62] If, as some suggest, there are only three generations of fermions then the symmetry group becomes SU(3)⊗SU(2)⊗U(1)⊗U(3)⊗U(3)⊗U(4) with a U(3) Generation group and a U(3) Layer group.

[63] In flat 4-dimensional space-time we showed in previous work (Blaha (2015a) and earlier books) that the Reality group had the form as the Standard Model interactions: SU(2)⊗U(1)⊗SU(3). This result followed from the restriction of the coordinate transformation group to the complex Lorentz group for complex coordinate systems and to the real Lorentz group for real coordinate systems. In the case of complex 4-dimensional space complex general coordinate transformations lead to the requirement of a local U(4) Reality group that can transform the coordinates of a complex coordinate system to real-valued coordinates. This U(4) group is a subgroup of the set of complex general

$$U(x) = \exp[i \textstyle\sum_k \Phi_k(x)\tau_k] \equiv e^{i\theta(x)} \tag{11.1}$$

where the sum is for k = 1, …, 16, where the $\Phi_k(x)$ are complex functions of x, and where τ_k is the kth generator represented as a 4×4 matrix. Eq. 11.1 represents U(x) in 4-dimensional complex space as a 4×4 matrix. We will generalize it into a curved space equivalent.

11.2.1 Vierbein Form of the 4×4 Reality Group Transformations

In this subsection we will map 4×4 Reality group transformations (eq. 11.1) to the form of General Relativistic transformations using the vierbeins (*tetrads)* that we introduced in chapter 4..Then, we will construct Complex General Relativistic transformations, the complex curved space Reality group affine connection, and then their associated dynamical equations.

The *vierbein* formalism begins with the Equivalence Principle that allows us to define an inertial coordinate system in the neighborhood of any point Z in space-time. We will use the notation $\varsigma^\alpha(Z)$ to denote the inertial coordinates at Z. We define a vierbein as

$$l^\alpha{}_\mu(x) = (\partial\varsigma^\alpha(x)/\partial x^\mu)_{x=Z} \tag{11.2}$$

In a neighborhood of Z, we can invert the relation between ς and x to define an inverse

$$w^\mu{}_\alpha(x) = (\partial x^\mu(\varsigma)/\partial\varsigma^\alpha)_{x=X} \tag{11.3}$$

such that

$$w^\mu{}_\alpha(x)l^\alpha{}_\nu(x) = \delta^\mu{}_\nu \tag{11.4}$$
$$w^\mu{}_\beta(x)l^\alpha{}_\mu(x) = \delta^\alpha{}_\beta \tag{11.5}$$

In real General Relativity all *vierbeins* are real-valued. In Complex General Relativity a *vierbein* $l^\alpha{}_\mu(x)$ is complex-valued.

The metric at a curved space-time point X is defined in terms of *vierbeins* as

$$g_{\rho\sigma}(x) = \eta_{\alpha\beta}\, l^\alpha{}_\rho(x)l^\beta{}_\sigma(x) \tag{11.6}$$
$$g^{\rho\sigma}(x) = \eta^{\alpha\beta}\, w^\rho{}_\alpha(x)w^\sigma{}_\beta(x) \tag{11.7}$$

The inverse of a *vierbein* transformation can also be expressed as

$$w_\beta{}^\nu(x) = l_\beta{}^\nu(x) = \eta_{\beta\alpha}g^{\nu\mu}(x)l^\alpha{}_\mu(x) \tag{11.8}$$

Then a *vierbein* and its inverse satisfy the relations

---

coordinate transformations and can be expressed in a manner reflecting the curvature of space with an affine connection. *Much of this section is a slightly revised form of material in Blaha (2015b).*

$$l^{\alpha}{}_{\mu}(x) l_{\beta}{}^{\mu}(x) = \delta^{\alpha}{}_{\beta} \tag{11.9}$$

and

$$l^{\alpha}{}_{\mu}(x) l_{\alpha}{}^{\nu}(x) = \delta^{\nu}{}_{\mu} \tag{11.10}$$

11.2.2 Transformations of Vierbeins

There are two general types of space-time transformations that can be performed on a vierbein.

1. A complex-valued (possibly real-valued) General Relativistic coordinate transformation:

$$l'^{\alpha}{}_{\mu}(x) = \partial x^{\nu}/\partial x'^{\mu}\, l^{\alpha}{}_{\nu}(x) \tag{11.11}$$

2. A complex-valued, local *Lorentzian transformation*

$$l'^{\beta}{}_{\mu}(x) = \Lambda(x)^{\beta}{}_{\alpha}\, l^{\alpha}{}_{\mu}(x) \tag{11.12}$$

where $\Lambda(x)^{\beta}{}_{\alpha}$ is an element of a subset of the local Complex Lorentz Group to be specified later.

11.2.3 Local Lorentzian Formalism for Vierbeins

The local Lorentzian transformations $\Lambda(x)^{\beta}{}_{\alpha}$ consist of local Lorentz transformations that are real-valued, and also complex-valued Lorentz transformations. Both types of transformations satisfy the orthogonality condition:

$$\eta_{\alpha\beta}\Lambda^{\alpha}{}_{\rho}(x)\Lambda^{\beta}{}_{\sigma}(x) = \eta_{\rho\sigma} \tag{11.13}$$

Thus the *vierbein* partakes of both local (position dependent) General Relativistic transformations and local Lorentzian transformations.

11.2.4 Vierbein Form of the 4×4 Reality Group General Coordinate Transformations

In eq. 11.1 we defined Reality group matrix transformations. Using the matrix form of this definition, and vierbeins, we see that we can express a Reality transformation in an inertial coordinate system in the neighborhood of a point Z in space-time. Thus

$$\begin{aligned} U^{\mu}{}_{\nu}(x) &\equiv w^{\mu}{}_{a}(x) U^{a}{}_{b}(x) l^{b}{}_{\nu}(x) = w^{\mu}{}_{a}(x)[e^{i\theta(x)}]^{a}{}_{b} l^{b}{}_{\nu}(x) \\ &= w^{\mu}{}_{a}(x)[\textstyle\sum_{n} (i\theta(x))^{n}/n!]^{a}{}_{b} l^{b}{}_{\nu}(x) \\ &= w^{\mu}{}_{a}(x) l^{a}{}_{\nu}(x) + w^{\mu}{}_{a}(x) i\theta(x)^{a}{}_{b} l^{b}{}_{\nu}(x) + \ldots \\ &= \delta^{\mu}{}_{\nu} + [\theta(x)]^{\mu}{}_{\nu} + [\theta(x)]^{\mu}{}_{\alpha}[\theta(x)]^{\alpha}{}_{\nu}/2 + \ldots \end{aligned} \tag{11.14}$$

where we have transformed the terms within the expansion of the exponentiated matrix into local Lorentz frame tensors, and where

$$[\theta(x)]^{\mu}{}_{\alpha} = w^{\mu}{}_{a}(x)\,[\theta(x)]^{a}{}_{b}\,l^{b}{}_{\alpha}(x) \qquad (11.15)$$

with a and b being the 4×4 matrix column and row indices of $[\tau_k]^a{}_b$ (See eq. 11.1.). The local Lorentzian tensorial form of U(x) can be used to develop Complex General Coordinate transformations as we do in the next section.

For later use we note that the inverse of eq. 11.15 is

$$U^{-1\alpha}{}_{\mu}(x) = w^{\alpha}{}_{a}(x)U^{-1a}{}_{b}(x)l^{b}{}_{\mu}(x) \qquad (11.16)$$

with

$$U^{-1\alpha}{}_{\mu}(x)U^{\mu}{}_{\nu}(x) = \delta^{\alpha}{}_{\nu} \qquad (11.17)$$

where

$$U^{-1a}{}_{b}(x) = [U^{\dagger}(x)]^{a}{}_{b} \qquad (11.18)$$

with † signifying hermitean conjugate.

11.2.5 Structure of Complex General Coordinate Transformations

Complex General Coordinate transformations can be uniquely factored into products of two terms. They have the form

$$\partial x''^{\nu}(x)/\partial x^{\mu} = U(x'')^{\nu}{}_{\beta}\,\partial x'^{\beta}(x)/\partial x^{\mu} \qquad (11.19)$$

where

$$x''^{\nu}(x) = U(x'')^{\nu}{}_{\beta}x'^{\beta}$$
$$x'^{\mu}(x) = U^{-1\mu}{}_{b}(x'')\,x''^{b}$$

where $U(x')^{\nu}{}_{\beta}$ is a complex coordinate transformation, and where $\partial x'^{\beta}(x)/\partial x^{\mu}$ is a purely real General Coordinate transformation.

We define

$$U(x'')^{\mu}{}_{\nu} = w^{\mu}{}_{a}(x'')\big[\exp(i\textstyle\sum_k \Phi_k(x'')\tau_k)\big]^{a}{}_{b}\,l^{b}{}_{\nu}(x'') \qquad (11.20)$$

$$U^{-1}(x'')^{\mu}{}_{\nu} = w^{\mu}{}_{a}(x'')\big[\exp(-i\textstyle\sum_k \Phi_k(x'')\tau_k)\big]^{a}{}_{b}\,l^{b}{}_{\nu}(x'') \qquad (11.21)$$

The uniqueness of the factorization follows from the Reality group (and U(4)) property that any complex 4-vector can be uniquely mapped to any specified real 4-vector.

Given the factorization (eq. 11.19) it becomes possible to separate the affine connection correspondingly.

11.2.6 Complex Affine Connection

The structure of a complex general coordinate transformation (eq. 11.19) enables us to calculate its affine connection for later use in determining the covariant derivative, and the dynamic equations. First the general coordinate transformation to complex or real valued coordinates x' from real-valued inertial coordinates ς^ρ is

$$\Gamma^{\sigma}{}_{\lambda\mu}(x') = \partial x'^{\sigma}/\partial\varsigma^{\rho}\, \partial^2\varsigma^{\rho}/\partial x'^{\lambda}\partial x'^{\mu} \qquad (11.22)$$

Next, a Reality group transformation from complex to real-valued coordinates when expressed as a transformation to coordinates x" from real-valued inertial coordinates ς^ρ has the affine connection

$$\Gamma_R{}^{\sigma}{}_{\lambda\mu}(x'') = \partial x''^{\sigma}/\partial\varsigma^{\rho}\, \partial^2\varsigma^{\rho}/\partial x''^{\lambda}\partial x''^{\mu} \qquad (11.23)$$

$\Gamma_R{}^{\sigma}{}_{\lambda\mu}(x'')$ can be rewritten as

$$\Gamma_R{}^{\sigma}{}_{\lambda\mu}(x'') = \partial x''^{\sigma}/\partial x'^{\beta}\, \partial x'^{\beta}(\varsigma)/\partial\varsigma^{\rho}\, \partial/\partial x''^{\mu}[\partial\varsigma^{\rho}/\partial x'^{\alpha}\, \partial x'^{\alpha}/\partial x''^{\lambda}] \qquad (11.24)$$

Using eq. 11.22 we find eq. 11.24 has the form

$$\Gamma_R{}^{\sigma}{}_{\lambda\mu}(x'') = \partial x''^{\sigma}/\partial x'^{\beta}\, \partial x'^{\alpha}/\partial x''^{\lambda}\, \partial x'^{\gamma}/\partial x''^{\mu}\, \Gamma^{\beta}{}_{\alpha\gamma}(x') + \partial x''^{\sigma}/\partial x'^{\beta}\, \partial^2 x'^{\beta}/\partial x''^{\lambda}\partial x''^{\mu} \qquad (11.25)$$

Next substituting the Reality group transformation

$$x''^{\nu}(x) = U(x'')^{\nu}{}_{\beta}x'^{\beta} \qquad (11.26)$$
$$x'^{\mu}(x) = U^{-1}(x'')^{\mu}{}_{\beta}\, x''^{\beta}$$

and using

$$\partial x''^{\sigma}/\partial x'^{\beta} = \partial[U(x'')^{\sigma}{}_{\alpha}x''^{\alpha}]/\partial x'^{\beta} = U(x'')^{\sigma}{}_{\beta} + x''^{\alpha}\,\partial U(x'')^{\sigma}{}_{\alpha}/\partial x'^{\beta} \qquad (11.27)$$
$$\partial x'^{\sigma}/\partial x''^{\beta} = \partial[U^{-1}(x'')^{\sigma}{}_{\alpha}x''^{\alpha}]/\partial x''^{\beta} = U^{-1}(x'')^{\sigma}{}_{\beta} + x''^{\alpha}\,\partial U^{-1}(x'')^{\sigma}{}_{\alpha}/\partial x''^{\beta} \qquad (11.28)$$

we find the second term in eq. 11.25 is the Reality group affine connection[64] from complex-valued coordinates x' to real-valued coordinates x"

$$\Gamma_R{}^{\sigma}{}_{\lambda\mu}(x'') = \partial[U(x'')^{\sigma}{}_{\alpha}x'^{\alpha}]/\partial x'^{\beta}\, \partial\{\partial[U^{-1}(x'')^{\beta}{}_{\alpha}x''^{\alpha}]/\partial x''^{\lambda}\}/\partial x''^{\mu} \qquad (11.29)$$

[64] Note the seeming contradiction of eqs. 11.25 and 11.29 reflect ‘differing’ affine conections between different pairs of coordinate systems.

11.2.7 Complex Curvature Tensor and Complex Einstein Equation of the 'New' Complex General Relativity

The 'new' complex space-time Riemann-Christoffel curvature tensor is

$$R^{\rho}{}_{\mu\nu\sigma}(x'') \equiv \partial\Gamma^{\rho}{}_{\mu\nu}(x'')/\partial x''^{\sigma} - \partial\Gamma^{\rho}{}_{\mu\sigma}(x'')/\partial x''^{\nu} + \Gamma^{a}{}_{\mu\nu}(x'')\Gamma^{\rho}{}_{\sigma a}(x'') - \Gamma^{a}{}_{\mu\sigma}(x'')\Gamma^{\rho}{}_{\nu a}(x'') \quad (11.30)$$

as can be shown by following the standard steps of its derivation. (We note that the algebra of the tensor manipulations is the same as in the usual derivation of General Relativistic quantities and equations.)

The Complex General Relativistic Einstein Equation is

$$R_{\mu\nu}(x'') - \tfrac{1}{2}\, g_{\mu\nu}R(x'') = -8\pi G\, T_{\mu\nu} \quad (11.31)$$

where G is Newton's gravitational constant ($5.674 \times 10^{-11}\ m^3kg^{-1}s^{-2}$), $T_{\mu\nu}$ is the energy-momentum tensor, the Ricci tensor is

$$R_{\mu\nu}(x'') = R^{a}{}_{\mu a\nu}(x'') \quad (11.32)$$

and the curvature scalar is

$$R(x'') = g^{\mu\nu}R_{\mu\nu}(x'') \quad (11.33)$$

with $g^{\mu\nu} = g^{\mu\nu}(x'')$.

11.2.8 'Infinitesimal' Form of the Affine Connections

Since the form of $\Gamma^{\sigma}{}_{\lambda\mu}(x'')$ as given in eqs. 11.25 – 11.29 is sufficiently complicated to make a solution of the new Einstein equation eq. 11.31 impossible currently, we will consider the infinitesimal form of the General Coordinate Reality Group affine connection. If the transformation from a real-valued coordinate system to a complex-valued coordinate system is 'infinitesimal' so that we can approximate the transformation with leading order terms then we can approximate $U(x'')^{\nu}{}_{\beta}$ and $U^{-1}(x'')^{\nu}{}_{\beta}$ of eqs. 11.20-11.21 with

$$U(x'')^{\nu}{}_{\beta} \approx \delta^{\nu}{}_{\beta} + i\sum_k \Phi_k(x'')[\tau_k]^{\nu}{}_{\beta} \quad (11.34)$$

$$U^{-1}(x'')^{\nu}{}_{\beta} \approx \delta^{\nu}{}_{\beta} - i\sum_k \Phi_k(x'')[\tau_k]^{\nu}{}_{\beta} \quad (11.35)$$

using

$$w^{\mu}{}_{a}(x'') \approx \delta^{\mu}{}_{a}$$

$$1^{b}{}_{\nu}(x'') \approx \delta^{b}{}_{\nu}$$

Then eq. 11.25 becomes

$$\Gamma_{tot}{}^{\sigma}{}_{\lambda\mu}(x'') \approx \Gamma_{GR}{}^{\sigma}{}_{\lambda\mu}(x') + \Gamma_{R}{}^{\sigma}{}_{\lambda\mu}(x'') \quad (11.36)$$

to leading order where $\Gamma_{GR}{}^{\sigma}{}_{\lambda\mu}(x')$ is a real-valued general relativistic affine connection and where

$$\Gamma_R{}^{\sigma}{}_{\lambda\mu}(x'') \equiv U(x'')^{\sigma}{}_{\beta}\, \partial[U^{-1}(x'')^{\beta}{}_{\lambda}]/\partial x''^{\mu} \approx -\tfrac{1}{2} i\{\textstyle\sum_k \partial\Phi_k(x'')/\partial x''^{\mu}\, [\tau_k]^{\sigma}{}_{\lambda} + \sum_k \partial\Phi_k(x'')/\partial x''^{\lambda}[\tau_k]^{\sigma}{}_{\mu}\} \tag{11.37}$$

with $\lambda\mu$ symmetry. Then

$$\Gamma_R{}^{\sigma}{}_{\lambda\mu} = -\tfrac{1}{2}i\{\textstyle\sum_k \partial\Phi_k(x'')/\partial x''^{\mu}\, [\tau_k]^{\sigma}{}_{\lambda} + \sum_k \partial\Phi_k(x'')/\partial x''^{\lambda}[\tau_k]^{\sigma}{}_{\mu}\}$$
$$\Gamma_R{}^{\sigma}{}_{\lambda\mu} = -\tfrac{1}{2}i\{\partial/\partial x''^{\mu}\, \varphi^{\sigma}{}_{\lambda} + \partial/\partial x''^{\lambda}\, \varphi^{\sigma}{}_{\mu}\} \tag{11.38}$$

with

$$\varphi^{\sigma}{}_{\mu} = \textstyle\sum_k \Phi_k(x'')[\tau_k]^{\sigma}{}_{\mu} \tag{11.39}$$

If we let

$$\Phi_k(x) = \int^{x} dy_{\lambda}\, A_{Rk}{}^{\lambda}(y) \tag{11.40}$$

then

$$\Gamma_R{}^{\sigma}{}_{\lambda\mu} = -\tfrac{1}{2}i\{\textstyle\sum_k A_{Rk}(x'')_{\mu}[\tau_k]^{\sigma}{}_{\lambda} + \sum_k A_{Rk}(x'')_{\lambda}[\tau_k]^{\sigma}{}_{\mu}\} = A_R{}^{\sigma}{}_{\mu\lambda} + A_R{}^{\sigma}{}_{\lambda\mu} \tag{11.41}$$

(summed over k) where the matrix $A_R{}^{\sigma}{}_{\mu\lambda}$ is given by

$$A_R{}^{\sigma}{}_{\mu\lambda} = -\tfrac{1}{2}i\textstyle\sum_k A_{Rk\mu}[\tau_k]^{\sigma}{}_{\lambda} \tag{11.42}$$

Note that τ_k has Lorentz indice[65]s due to eqs. 11.14 and 11.15. We have established the U(4) Reality group in a local frame and then extended it to complex curved space using complex vierbeins.

11.2.9 Form of the General Relativistic Reality Group Interaction

Also note eq. 11.41 has a form similar to eqs. 4.23, 4.27, and 4.32. We will now extend the formulation to a two field formalism as in chapter 4 by defining the terms[66] $A_R{}^{1\sigma}{}_{\mu\lambda}$ and $A_R{}^{2\sigma}{}_{\mu\lambda}$ which yields

$$\Gamma_R{}^{1\sigma}{}_{\mu\lambda} = A_R{}^{1\sigma}{}_{\mu\lambda} + A_R{}^{1\sigma}{}_{\lambda\mu}$$
$$\Gamma_R{}^{2\sigma}{}_{\mu\lambda} = A_R{}^{2\sigma}{}_{\mu\lambda} + A_R{}^{2\sigma}{}_{\lambda\mu} \tag{11.43}$$

---

[65] In a local Lorentz frame, and in flat complex or real space, the Lorentz indices become U(4) matrix indices since the vierbein factors become Kronecker δ-functions.

[66] We use two affine connections with corresponding gauge fields here in order to be able to take advantage of the canonical framework for deriving dyamical equations from the lagrangian resulting from the use of the augmented Riemann-Christoffel curvature tensor to generate lagrangian terms.

in 'addition'[67] to the 'real-valued' affine connections $\Gamma_{GR}{}^{1\sigma}{}_{\lambda\mu}$ and $\Gamma_{GR}{}^{2\sigma}{}_{\lambda\mu}$ in the expression for the generalized Riemann-Christoffel curvature tensor specified later in section 11.5. The introduction of two Reality affine connections, and gauge fields, $A_R{}^j{}_\mu$

$$A_R{}^{j\sigma}{}_{\mu\lambda} = -\tfrac{1}{2}i\sum_k A_R{}^j{}_{k\mu}[\tau_k]^\sigma{}_\lambda \tag{11.44}$$

for j = 1, 2, has the same motivation as the introduction of secondary fields and affine connections in chapter 4. Under a local U(4) gauge transformation C_R *in flat space-time* with

$$A_R{}^{j\mu}{}_{flat} = -\tfrac{1}{2}i\sum_k A_{Rflat}{}^j{}_k{}^\mu\tau_k \tag{11.45}$$

we find

$$\begin{aligned} A_{Rflat}{}^{1\mu}(x) &\rightarrow C_R(x)A_{Rflat}{}^{1\mu}(x)C_R{}^{-1}(x) - i\,C_R(x)\partial^\mu C_R{}^{-1}(x) \\ A_{Rflat}{}^{2\mu}(x) &\rightarrow C_R(x)A_{Rflat}{}^{2\mu}(x)C_R{}^{-1}(x) \end{aligned} \tag{11.46}$$

We now see that the derivation of the Riemann-Christoffel curvature tensor which used

$$H^\sigma{}_{\nu\mu} = \Gamma^\sigma{}_{\nu\mu} + \Gamma^{2\sigma}{}_{\nu\mu} \tag{4.53}$$

should extend $H^\sigma{}_{\nu\mu}$ for the case of real-valued coordinates to

$$H^\sigma{}_{\nu\mu} = \Gamma_{GR}{}^\sigma{}_{\nu\mu} + \Gamma_{GR}{}^{2\sigma}{}_{\nu\mu} + \Gamma_R{}^{1\sigma}{}_{\nu\mu} + \Gamma_R{}^{2\sigma}{}_{\nu\mu} \tag{11.47}$$

In flat space the affine connections $\Gamma_R{}^{i\sigma}{}_{\mu\lambda}$ become a sum of gauge fields by eq. 11.42 with

$$A_{Rflat}{}^i{}_\mu{}^a{}_b = A_{Rflat}{}^i{}_\mu[\tau_k]^\sigma{}_\lambda\delta_\sigma{}^a\delta^\lambda{}_b \tag{11.48}$$

$A_{Rflat}{}^{ia}{}_{\mu b}$ can be expressed in matrix form as

$$A_{Rflat}{}^i{}_\mu = -\tfrac{1}{2}i\sum_k A_{Rflat}{}^i{}_{k\mu}\tau_k \tag{11.49}$$

since the vierbeins used to define eq. 11.42 become flat space-time metric tensors with the flat space vierbein: $l^\lambda{}_b = \delta^\lambda{}_b$.

---

[67] Actually we are expanding the affine connection $\Gamma^\sigma{}_{\lambda\mu}$ of chapter four into a real and a Reality group part.

11.2.10 Distinction Between the General Coordinate Reality Group and the Special Relativistic Reality Group

The General Coordinate Reality Group described in this section differs from the Special Relativistic Reality Group described in Blaha (2015a) and earlier books. The difference is due to the requirement of Complex Special Relativity to satisfy the Lorentz condition

$$\Lambda(\omega, \mathbf{u})^T G \Lambda(\omega, \mathbf{u}) = G \tag{11.50}$$

for any real or complex Lorentz transformation $\Lambda(\omega, \mathbf{u})$ where $G = \text{diag}(1, -1, -1, -1)$. (See appendix 4-A of Blaha (2015a) for more details.)

Complex General Coordinate transformations are not required to satisfy the above condition. Thus $\Gamma_{GR}{}^{\sigma}{}_{\nu\mu}$ and $\Gamma_R{}^{\sigma}{}_{\mu\lambda}$ are independent, and separate, and have separate Reality groups. The Unified Standard Model group is SU(3)⊗SU(2)⊗U(1)⊗U(4)⊗U(4)⊗U(4) with both an SU(3)⊗SU(2)⊗U(1) part due to complex Special Relativity, and a U(4) group part due to complex General Relativity.

The the set of all General Coordinate Reality group transformations can be separated into two disjoint sets: 1) the set of transformations that satisfy eq. 11.50 (and thus yield SU(3)⊗SU(2)⊗U(1) symmetry when combined with the set of real-valued coordinate transformations and taken to the flat space-time limit), and 2) the set of transformations that do *not* satisfy eq. 11.50 (and thus have U(4) symmetry). These sets do not overlap and each has its own symmetry group.

Transformations in each set have a 'multiplication' rule such that the product of two elements of the set is in the set. Thus each set consists of the elements of a group of transformations. The SU(3)⊗SU(2)⊗U(1) group structure of the transformations of the first set follows directly from eq. 11.50 as we showed in Blaha (2015a) and earlier books. The U(4) property of the second set follows from the defining characteristic of U(4) – any 4×4 complex matrix can be made a real-valued matrix using U(4). The real-valued matrix (transformation) so produced does not satisfy eq. 11.50.

11.2.11 Spontaneous Reality Group Symmetry Breaking

As we showed in Blaha (2015b) the Reality group for complex General Relativity is subject to spontaneous symmetry breaking. There are three important consequences:

1. The transition from real to complex space is 'prevented' by the symmetry breaking. This feature accounts for our presence in real space-time. Direct evidence for the complex nature of space-time is not possible due to the symmetry breaking.

2. The symmetry breaking leads to the generation of mass contributions to Reality group gauge fields and to all fermions. Thus Reality group gauge field interactions are severely reduced due to presumably ultra-large vector boson masses (perhaps of the

order of the Planck mass) and due to an ultra-weak coupling constant. Thus complex General Relativistic Reality group interactions would be hard to detect.

3. The fermion mass contributions (to all fermions) set the mass scale for fermions and thus account for the equality of gravitational mass and inertial mass. See Blaha (2015b) for details, which are partly presented in the following extract for the reader's convenience.

4. Note that the focus of this book is primarily on the particle interactions part of the Theory of Everything lagrangian. The interactions presented here will appear in the fermion sector of The Theory of Everything lagrangian through their appearance in the total covariant derivative seen in chapter 4 and again, with additions, in section 11.5.

We note that Blaha (2015b) differs in some details with the discussion of this section. The primary difference is our introduction of Reality group gauge fields here. Below is an extract from Blaha (2015b) relating to the above items that are reprinted for the reader's convenience.

EXTRACTS FROM Blaha (2015b) and (2016c)

3.1 The Enigma of Higgs Particles and the Higgs Mechanism

In our previous work on the Standard Model, and its generalization to The Extended Standard Model described in a series of books entitled *Physics is Logic*, we showed that the fermion spectrum results from Complex Special Relativity, the gauge interactions result from the Reality group, the fermion generations result from the Generation group, the layers of fermions result from the U(4) Layer group, and from the combination with Complex General Relativity in our Theory of Everything. Higgs particles and the Higgs Mechanism were inserted *ad hoc* to generate particle masses and symmetry breaking effects.

The apparent recent discovery of Higgs particles at CERN seems to solidify the existence of the Higgs sector of the Standard Model and of our Extended Standard Model as described in earlier volumes of *Physics is Logic*.[68]

But whence arises Higgs particles? There does not appear to be a more fundamental cause than the need for particle masses obtained through symmetry breaking. And so the Higgs sector was an expedient mechanism. With our method of avoiding divergences in perturbation theory using Two-Tier quantum field theory the need for the Higgs Mechanism appears to have disappeared with the former need for a symmetry breaking mechanism to generate particle masses. The ElectroWeak sector has no divergences in our approach and thus does not need the renormalization program previously developed that was based on symmetry breaking using the Higgs Mechanism.

[68] Blaha (2015a) and (2015b).

In chapter 4 we will see that Higgs fields naturally appear for gauge fields that can be made real-valued, and do not appear for one set of gauge fields that cannot be made real-valued: the Strong Interaction SU(3) gauge fields.

In considering the Higgs Mechanism a number of peculiarities appear that diminishes its attractiveness:

1. As remarked above, it is selective in the sense that some gauge fields have associated Higgs particles and utilize the Higgs Mechanism, and some gauge fields do not have associated Higgs particles. In particular, the ElectroWeak gauge fields, the Generation group gauge fields, the Layer group gauge fields, and the complex gravitation fields have associated Higgs particles. The strong interaction (gluon) gauge fields do not. See chapter 4 for the reason.

2. The conventional Higgs potentials have a quadratic mass term of the "wrong" sign plus a quartic interaction term, which together, generate non-zero vacuum expectation values. They obviously accomplish their goal. But the source of these potentials, and why they have their form, is unknown. One suspects a fundamental principle should be operative here.

3. One can imagine creating a Higgs microscope at some super-accelerator. Using this microscope in the presence of a (classical) condensate could enable the Uncertainty Principle to be violated. This possibility, in the case of a microscope using electromagnetic fields, was the source of a heuristic argument for the need to quantize the electromagnetic field.[69]

4. The standard formulation of the Higgs Mechanism uses classical fields under the assumption that a path integral formulation justifies their use. While this may be true, the path integral formulation relies on implicit, unstated boundary conditions that obscure the physics of the quantum field theoretic nature of the mechanism. A direct quantum field theoretic study of the Higgs Mechanism is needed and would further elucidate its character. It is possible, and it has been shown in our earlier books, that the apparently "true" mechanism described below reveals a number of important new results in a properly formulated version of the Higgs Mechanism.

3.2. "True" Origin of an Acceptable Mass Creation Mechanism

In this chapter we will describe a new mechanism that utilizes an extension of quantum field theory to include classical fields that we have called *pseudoquantization*[70] and *pseudoquantum field theory.* It combines both quantum and classical fields within the same

[69] Heitler (1954) p. 86 provides a good discussion of the need to quantize the electromagnetic field.
[70] This new formalism was first described in S. Blaha, Phys. Rev. D**17**, 994 (1978).

framework. In this extended theory vacuum expectation values appear as coherent ground states that are strictly classical in nature.

This chapter is based on our 1978 paper that appeared in *Physical Review D.* We suggest the reader skim or read the paper before proceeding, or refer to the preceding chapter for details as necessary while reading this chapter. The paper also presents a new formulation of Quantum Mechanics that incorporates both quantum and classical mechanics within one framework that is of interest in its own right. Recently, experimenters have been investigating the possibility of macroscopic and other strange quantum phenomena. The new formulation is ideally suited for tracing the transition from a quantum to a classical regime. For example, it is applicable to "large n atoms" where the outermost electrons approach classical behavior with an almost continuous energy spectrum.

3.3 Higgs-Like Vacuum Expectation Value Generation of Masses

The Higgs Mechanism is based on the appearance of non-zero c-number vacuum expectation values for Higgs fields due to potential terms directly appearing in lagrangians.

3.3.1 Pseudoquantization of Higgs Particles

We will now consider the pseudoquantization of a scalar particle using two fields in a manner shown in the preceding chapter. It will become a "Higgs" particle with a non-zero vacuum expectation value.

Using the formalism of the preceding chapter we define $\varphi_1(x)$ and $\varphi_2(x)$[71] for a generic boson suppressing any internal symmetry indices for simplicity. We define a "vacuum state" containing a coherent superposition of the form of eq. 2.11 that satisfies

$$\varphi_1(x)|\Phi, \Pi> = \Phi|\Phi, \Pi> \quad (2.12)$$

where Φ is a constant. Evaluating a fermion interaction term we find a mass term emerges[72]

$$\overline{\psi}(\varphi_1 + \varphi_2)\psi \rightarrow \overline{\psi}(\Phi + \varphi_2)\psi \quad (2.13)$$

It can also generate a mass for an interaction with a gauge field of the form

$$A^{\mu}(\varphi_1 + \varphi_2)^2 A_{\mu} \rightarrow A^{\mu}(\Phi + \varphi_2)^2 A_{\mu} \quad (2.14)$$

for ElectroWeak and other gauge fields. The φ_2 term leads to the production of Higgs particles in interactions. (The production of Higgs particles that decay into ElectroWeak gauge particles has recently been found at CERN.)

---

[71] The subscripts on the fields are not gauge symmetry indices but simply identifiers distinguishing the fields from one another.

[72] When matrix elements with a "vacuum state" such as eq. 2.12 are taken.

The present formalism provides a clean way to separate the vacuum expectation value of a scalar particle from its quantum field part in contrast to the conventional Higgs Mechanism where one has to separate a Higgs field into parts manually.

To obtain both the vacuum expectation value and the interaction with the quantum part of the pseudoquantum fields we choose to always specify interactions with fermions and gauge fields using $\varphi = \varphi_1 + \varphi_2$ as seen above.

It appears that our formulation of the mass generation mechanism sheds significant light on the reason for the special prominence of inertial frames. Consider massive scalars.[73] Eqs. 2.12 can describe a massive scalar particle. If the scalar is massive, then the rest frame particle "vacuum" coherent state below yields a non-zero expectation value Φ:

$$|\Phi, \Pi\rangle = C\exp\{[(2\pi)^3 m/2]^{1/2}\Phi[a_2^\dagger(\mathbf{0},m) + a_2(\mathbf{0},m)]\}|0\rangle \qquad (3.1)$$

where m is a generic mass. (We note that the conventional Higgs Mechanism also has mass terms.) *Thus our pseudoquantum formalism allows us to define coherent "vacuum" states that lead to particle masses and Higgs particles.*

3.4 Why Inertial Reference Frames are Special

The great physicists of the early 20th century raised numerous questions about Special Relativity after Einstein and Poincarè's discovery. Prominent among them was the question of why inertial reference frames are of especial importance in Special Relativity, and afterwards in General Relativity.

It appears that our formulation of the mass generation mechanism sheds significant light on the reason for the special prominence of inertial frames. Earlier we considered the case of a massless pseudoquantum scalar. We now consider massive scalars since experiments at CERN have apparently discovered a Higgs particle with a 125 GeV/c mass. The above equations describe a massive scalar particle. If the scalar is massive, then a "vacuum" coherent state, which yields a non-zero expectation value, exists for a particle of mass m in its rest frame.

Then, having established a preferred frame for a Higgs particle, in The Extended Standard Model, and requiring that invariant intervals

$$ds^2 = dt^2 - d\mathbf{x}^2 \quad \text{(in rectangular coordinates)} \qquad (3.2)$$

are unchanged by a (complex or real) Lorent transformation, we find that inertial reference frames are singled out as "special" in the sense that they are the only accessible reference frames that can be generated by a Lorentz boost/transformation from a Higgs particle rest frame. *The Higgs particle vacuum state singles out the class of inertial reference frames.*

Thus Higgs particles play a central role in establishing the basis of physical reality.

---

[73] Experiments at CERN have apparently discovered a Higgs particle with a 125 GeV/c mass.

3.5 T Invariance of Our Pseudoquantum Scalar Particle Theory

Pseudoquantum scalar particle hamiltonian equations are invariant under time reversal: $t \rightarrow t' = -t$. The vacuum states defined by eqs. 2.1 – 2.8 break the time reversal invariance of the scalar pseudoquantum theory resulting in retarded particle propagators.

The hamiltonian equations

$$[H, \varphi_1(\mathbf{x}, t)] = -i\partial\varphi_1/\partial t \qquad (3.3)$$
$$[H, \varphi_2(\mathbf{x}, t)] = -i\partial\varphi_2/\partial t$$

are invariant under time reversal. If we define a time reversal operator transformation U then the time-reversed equations are

$$[UHU^{-1}, \varphi_1(\mathbf{x}, -t)] = +i\partial\varphi_1(\mathbf{x}, -t)/\partial(-t) \qquad (3.4)$$
$$[UHU^{-1}, \varphi_2(\mathbf{x}, -t)] = +i\partial\varphi_2(\mathbf{x}, -t)/\partial (-t)$$

The operator U, which is unitary, transforms H into –H. This operation is legal because the hamiltonian – in this case – is not positive definite and admits negative energy states.[74] Thus

$$[H, \varphi_1(\mathbf{x}, -t)] = -i\partial\varphi_1(\mathbf{x}, -t)/\partial (-t) \qquad (3.5)$$
$$[H, \varphi_2(\mathbf{x}, -t)] = -i\partial\varphi_2(\mathbf{x}, -t)/\partial (-t)$$

and the time reversal invariance of the equations of motion is established for this case.

Time reversal invariance is broken by our choice of vacuum states. This choice is necessary to obtain classical field states as we showed in the preceding chapter. A demonstration of time reversal symmetry breaking is presented in the following chapter where we show the theory has retarded propagators for particle propagation in "in" and "out" asymptotic states.

Within the interaction region the particle propagators are the sum of retarded and advanced parts that combine to yield principal value propagators – not Feynman propagators.

Many years ago Feynman and Wheeler championed principal value propagators for electrodynamics to obtain an action-at-a-distance theory of Quantum Electrodynamics. While their theory and ours differ from the standard quantum field theory approach there is no reason to view them as faulty, or having serious physical defects. The only question is whether nature chooses conventional quantum field theory or pseudoquantum quantum field theory. In our case the need for classical scalar particle non-zero vacuum expectation values strongly motivates our choice of psedoquantum Higgs particles.

[74] Unlike the usual case of second quantized Klein-Gordon quantum field theory.

3.6 Retarded Propagators for Our Pseudoquantum Higgs Particles

In the previous chapter we pointed out that our pseudoquantization Higgs theory has an arrow of time due to its boundary conditions as expressed in its definition of the vacuum state and its dual. In this chapter we will show that the theory uses retarded propagators for propagation to and from the interaction region to asymptotic in-states and out-states. Within an interaction region the theory uses half-retarded – half-advanced propagators. The Appendix 2-A paper has a detailed discussion of propagators between eqs. 145 and 149, and also in section V. We will discuss some aspects of the perturbation theory and propagators of our scalar particles in this chapter.

First we note that in-states at $t = -\infty$ are composed of superpositions of $a_2(k)$ and $a_2^\dagger(k)$ creation and annihilation operators since

$$a_2(k)|0\rangle \neq 0 \qquad\qquad a_2^\dagger(k)|0\rangle \neq 0 \tag{3.6}$$

while out-states are composed of superpositions of $a_1(k)$ and $a_1^\dagger(k)$ creation and annihilation operators:

$$\langle 0|a_1(k) \neq 0 \qquad\qquad \langle 0|a_1^\dagger(k) \neq 0 \tag{3.7}$$

Consequently when in-state particles (x_1) propagate into the interaction region (x_2) the relevant propagators are retarded propagators with the form

$$\begin{aligned} G_{in}(x_2, x_1) &= \langle 0|T(\varphi_{1\,in}(x_2), \varphi_{2\,in}(x_1))|0\rangle \\ &= \theta(x_{20} - x_{10})\langle 0|[\varphi_{1\,in}(x_2), \varphi_{2\,in}(x_1)]\,|0\rangle \end{aligned} \tag{3.8}$$

by eq. 148 of Appendix 2-A. Eq. 3.8 is a manifestly retarded propagator. The choice of vacuums clearly results in a time asymmetry giving a retarded propagation reflecting the familiar Arrow of Time.

A similar situation prevails for propagation to out-states (x_3) from the interaction (x_2) region:

$$\begin{aligned} G_{out}(x_3, x_2) &= \langle 0|T(\varphi_{1\,out}(x_3), \varphi_{2\,out}(x_2))|0\rangle \\ &= \theta(x_{30} - x_{20})\langle 0|[\varphi_{1\,out}(x_3), \varphi_{2\,out}(x_2)]\,|0\rangle \end{aligned} \tag{3.9}$$

Within the interaction region the Higgs particles have principal value propagators as shown in section V of Appendix 2-A,

Thus we find pseudoquantum Higgs particles embody a local Arrow of Time. The locality of the Arrow of Time is embodied in all the particles that interact with the Higgs particle. Since the mass of *every* particle – bosons and fermions – has a Higgs contribution, and thus *every* particle[75] interacts with Higgs particles, the local Arrow of Time permeates The Extended Standard Model as well as the more familiar Standard Model known from experiment.

---

[75] Excepting photons.

The local Arrow of Time differs from the *macroscopic* Arrow of Time, which is statistical in nature.

3.7 The *Local* Arrow of Time

In the *Physics is Logic* monographs we saw that complex coordinates led to the form of the fermion spectrum, that the mapping of complex coordinates to real-valued coordinates yielded the Reality group and The Extended Standard Model gauge interaction, that Complex General Relativity led to Higgs particles that were directly united with elementary article masses and gave us the equality of inertial mass and gravitational mass, that the Layer group leads to four layers of fermions, and that the reduction of complex gauge fields to real gauge fields explained the appearance of Higgs fields in The Standard Model and The Extended Standard Model (chapter 4).

Now we see that the pseudoquantization procedure leads to retarded Higgs field propagators and thence to a *local* arrow of time. Many arguments have been put forward over the past hundred plus years for the Arrow of Time. Many arguments based on Statistical Mechanics, Entropy, and Boltzmann's statistical atomic theory have suggested the Arrow of Time is a global statistical consequence. This view seems to contradict the results of elementary particle experiments where a *local* Arrow of Time is evident. It appears that the Arrow of Time has two sources: local and global.

Our rationale for the Arrow of Time begins with retarded Higgs fields. Then we note that Higgs field quantum interactions appear for all fermions and gauge particles. Thus all particle interactions are imbued with an Arrow of Time. Particles that are united to form macroscopic matter inherit their combined local Arrows of Time producing a global Arrow of Time that we experience.

Thus our pseudoquantiztion approach offers a more satisfactory solution of the origin of the Arrow of Time.

It is remarkable that complex quantities – coordinates and fields – through the Higgs phenomena that we have considered, lead to the equality of inertial mass and gravitational mass (see below), and an Arrow of Time. This unity of mass and time phenomena may reflect the deeper fact that we can have no practical Arrow of Time if all particles were massless, for particle dynamics at light speed would then be pointless. This view has been expressed by DeWitt, Unruh, and others who have pointed out that, physically, time is meaningful and measurable only if masses exist; the larger the mass, the more accurate the time measurement in principle.[76]

...

[76] No mass, no clock; no clock, no physical time. See Blaha (2015a) pp. 368-371 for a discussion including comments by DeWitt and Unruh.

6. The Equality of Inertial Mass and Gravitational Mass

From the days of Newton through Einstein[77] to the present the equality of gravitational mass and inertial mass has been a topic of interest. Mach, who played an important role, in this ongoing discussion, thought distant masses in the universe were the source of the equality. However the origin of the equality, which has been shown experimentally to very high accuracy, remained uncertain until the present work where we show the interconnection of the Extended Standard Model and Complex Gravitation via Higgs generated masses unites gravitational and inertial mass.

In chapter 5 we showed that Complex General Relativity could be formulated in a manner similar to the Extended Standard Model in which the Reality group played a part. This formulation leads to scalar particles that can be viewed as Higgs particles since the fields could be shifted by a constant without affecting the kinetic part of their dynamic equations. If a Higgs potential is present then these fields could undergo spontaneous breakdown and then have non-zero vacuu expectation values.

The decision to base gravitational Higgs fields on the Reality group transformations of Complex General Relativity was based on a desire to build a Theory of Everything. (See chapter 7.) The present Complex Gravity theory, that we have developed, directly entwines gravitation and particle physics through the Reality group. The gravitational Higgs equations then become an elegant, compact unifying feature of a Theory of Everything. The known Higgs equations of the Extended Standard Model are now intertwined with the gravitational Higgs equations, which are a consequence of Complex General Relativity (suitably extended).[78] The gravitational Higgs potential energy-momentum contribution remains to be justified but can be provisionally inserted by hand just as Higgs potentials are inserted in The Extended Standard Model.

Since fermion field masses are now sums of ElectroWeak Higgs contributions, Generation group Higgs contributions, and gravitational Higgs contributions and since the gravitational Higgs fields appear in all fermion masses the equality of inertial and gravitational mass is proven. The gravitational Higgs particles equations depend, in part, on the gravitational field by eq. 5.50 and so set the mass scale of the gravitational mass. The presence of the gravitational Higgs contributions ***for all fermions****, sets the scale of the inertial Higgs field contributions from the Extended Standard Model Higgs particles.*

Since an expression cannot mix mass scales, the gravitational mass scale must be the same as the inertial mass scale. Inertial Mass equals gravitational mass.

We have established the equality of inertial and gravitational mass at the short distance quantum level. In our view, this explanation is far more satisfying than basing the equality on a

[77] For example, Einstein and Grossman in 1913 stated, "The theory herein described originates in the conviction that the proportionality between the inertial and gravitational mass of a body is an exact law of nature that must be expressed as a foundation principle of theoretical physics."

[78] This approach is further supported by the use of the Higgs mechanism to produce cosmic inflation and justify the expansion of the universe. See Guth and colleagues for discussions of Higgs induced inflation.

combination of large distance phenomena and quantum phenomena. As Einstein and Weyl have pointed out: all fundamental physics phenomena should be based on a local theory. Complex Gravity, as we have constructed it, combined with the Extended Standard Model furnishes a completely local basic Theory of Everything.

END OF ABSTRACT FROM Blaha (2015b) and (2016c)

11.3 Generation Group Interactions

The Generation group leads to the existence of four fermion generations for each of the four types of fermions and generates interactions between fermions and bosons. The features of this group have been described in earlier books. We will first present the description of the origin and features of the Generation group and then end the section with a discussion of the gauge field interaction terms that appear as a result in covariant derivatives and in the Riemann-Christoffel curvature tensor.

EXTRACTS FROM Blaha (2015a) and (2016c)

15. Baryon, Lepton, Dark Baryon, and Dark Lepton Conservation Laws and New U(4) Gauge Symmetry

15.1 Baryon Number Conservation and a Possible Baryonic Force

We have considered baryon number conservation and a possible baryonic force in Blaha (2014a) and (2014b). Much of this section contains selections from these books.

The primary forces involved in the interactions and collisions of baryons are the forces of The Standard Model, the force of gravity, and a fifth force which we take to be the baryonic force, a much discussed possible force that depends on the baryon numbers of objects experiencing it. The force (neglecting Standard Model interactions) between two clumps of baryonic matter containing baryons and other particles: clump1 being of mass m_1 and baryon number n_1, and clump2 being of mass m_2 and baryon number n_2 is

$$F = -Gm_1m_2/r^2 + (\beta^2/4\pi)n_1n_2/r^2 \tag{15.1}$$

where G is the gravitational constant, β is the baryonic constant and r is the distance between widely separated clumps. Experimentally a baryonic force between baryons has not been detected with any degree of certainty. Sakurai (1964) discusses early efforts in detail. Eőtvős experiments as far back as 1922 on the ratio of the observed gravitational mass to the inertial mass showed that that it is constant to within one part in 100,000,000 indicating the baryonic force, if it exists, as we believe it does, is extremely weak in comparison to the gravitational force. Eőtvős et al[79] found

$$(\beta^2/4\pi)/(Gm_p^2) < 10^{-5}$$

where m_p is the proton mass.

Since then, the experiment has been redone with improved accuracy by Dicke and collaborators.[80] They have improved the accuracy to one part in 100 billion. A further analysis showed a very small discrepancy that suggested the ratio, while small, was non-zero, implying the equivalence principle might not be exact and that the discrepancy changed with the material used in the experiment – just what one might expect if a very small baryonic force was present – often called the "fifth force." At present the existence and amount of the discrepancy is unclear. Nevertheless, we will assume a fifth force.

The primary rationale for the fifth force is the apparent conservation of baryon number. The conservation of baryon number has been repeatedly investigated by experimenters and found to be true to extremely high accuracy. For decades theorists have suggested that a baryon conservation law[81] follows from the existence of a gauge field in a manner much like electric charge conservation follows from the properties of the electromagnetic gauge field.

15.1.1 Estimate of the Baryonic Coupling Constant

The baryonic force, and coupling constant, is known to be very small in comparison to gravity and the other known forces. Measurements of the gravitational constant G are significantly different.[82,83] The reason(s) for these discrepancies is not known. We will assume that both the 2010 and 2013 measurements of G are experimentally correct but disagree because of the baryonic force term in eq. 15.1 that would create a difference in effective G values if the experiments used different masses and thus baryon numbers. Quinn et al found a value for the gravitational constant of $G_1 = 6.67545\times10^{-11}$ m^{3}kg^{-1}s^{-2}. The combined 2010 CODATA value for the gravitational constant was $G_2 = 6.67384\times10^{-11}$ m^{3}kg^{-1}s^{-2}. Both values are subject to estimated uncertainties.

Suppose these values are correct and are due to a difference in the chemical composition (metals) of the test masses used in the experiment. Quinn et all use 1.2 kg test masses composed of Cu-0.7% Te free machining alloy. The CODATA value being a composite of many experiments does not have an effective equivalent test mass value, or composition specified.[84] Suppose the test mass value is $N_1^2m_1^2 + N_{1e}^2m_e^2$ for the G_1 result giving

---

[79] Eőtvős, R. V., Pekár, D., Fekete, E., Ann. d. Physik **68**, 11 (1922).

[80] P. G. Roll, R. Krotkov, R. H. Dicke, Annals of Physics, 26, 442, 1964.

[81] See Gell-Mann, M. and Levy, M. *Nuovo Cimento* 16, 705 (1960) for a proof and Sakurai (1964) for a discussion of the relation of the baryonic gauge field to gravity experimentally.

[82] T. Quinn et al, Phys. Rev. Lett. **111**, 101102 (2013).

[83] P. J. Mohr, B.N. Taylor, and D. B. Newell, Rev. Mod. Phys. 84, 1527 (2012).

[84] The Eötvös' experiment used a 0.1 gm test mass of $RaBr_2$. R. v. Eötvös, D. Pekár, E. Fekete, Annalen der Physik (Leipzig) 68, 11, 1922.

$$-(N_1^2 m_1^2 + N_{1e}^2 m_e^2)G_1 = [-G(m_1^2 N_1^2 + N_{1e}^2 m_e^2) + (\beta^2/4\pi)N_1^2] \quad (15.2)$$

where G is the real value of the gravitational constant. The total test mass is $(m_1^2 N_1^2 + N_{1e}^2 m_e^2)$ with N_1 baryons of average mass m in each test mass and N_{1e} leptons of average mass m_e.

Suppose further the test mass value is $N_2^2 m_2^2 + N_{2e}^2 m_e^2$ for the G_2 result giving

$$-(N_2^2 m_2^2 + N_{2e}^2 m_e^2)G_2 = [-G(m_2^2 N_2^2 + N_{2e}^2 m_e^2) + (\beta^2/4\pi)N_2^2] \quad (15.3)$$

where G is the real value of the gravitational constant. The total test mass is $(m_2^2 N_2^2 + N_{2e}^2 m_e^2)$ with N_2 baryons of average mass m_2 in each test mass and N_{2e} leptons of average mass m_e. Since the test masses are electrically neutral and there are approximately equal numbers of protons and neutrons in a test mass it follows approximately that

$$N_{1e} = \tfrac{1}{2}N_1 \quad \text{and} \quad N_{2e} = \tfrac{1}{2}N_2 \quad (15.4)$$

Subtracting eq. 15.2 from eq. 15.3 after some algebra[85] we find

$$\begin{aligned}\Delta G = -G_2 + G_1 &= (\beta^2/4\pi)/(m_2^2 + m_e^2/2) - (\beta^2/4\pi)/(m_1^2 + m_e^2/2) \\ &\simeq (\beta^2/4\pi)(1/m_2^2 - 1/m_1^2)\end{aligned} \quad (15.5)$$

The masses m_1 and m_2 can differ. For example, if m_H is mass of the hydrogen atom, then $m^{-1} = 1.0m_H^{-1}$ for hydrogen, for carbon $m^{-1} = 1.00782m_H^{-1}$, for copper $m^{-1} = 1.00895m_H^{-1}$, and for lead $m^{-1} = 1.00794m_H^{-1}$.[86] Thus using the Quinn et al and CODATA results and assuming copper and lead test masses we find the order of magnitude *estimate*:

$$\begin{aligned}\alpha_B = \beta^2/4\pi &\simeq \Delta G/[(1.00895^2 - 1.00794^2)\, m_H^2] \\ &\simeq \Delta G/G\, G\, m_H^2/.002037 \\ &\simeq (0.000241/0.002037)Gm_H^2 \\ &\simeq .118\, Gm_H^2\end{aligned} \quad (15.6)$$

indicating a very weak baryonic force consistent with our general view of the Megaverse. The baryon fine structure constant is minute in comparison to the electromagnetic fine structure constant $\alpha \simeq 1/137$.

Due to our assumptions in the calculation of α_B, which makes it merely an order of magnitude estimate at best, we suggest that an experimental group measure G with differing test masses in the same apparatus to obtain a better value for α_B.

---

[85] The reduction of the calculation to algebra reminds the author of Nobelist Hans Bethe's remark that he only felt he understood a physical phenomenon when he could reduce it to algebra. This was quite evident when the author collaborated with Professor Bethe on a study of pion condensation in neutron stars some years ago.

[86] "One Hundred Years of the Eötvös Experiment", I. Bod, E. Fischbach, G. Marx and Maria Náray-Ziegler, August, 1990.

15.1.2 A Baryonic Gauge Field

Based on a reasonable value for α_B we assumed in Blaha (2014) that a baryonic gauge field exists that is similar to the electromagnetic field except for features due to its existence in the 16-dimensional universe that we called the Megaverse.[87] This gauge field couples extremely weakly[88] to individual baryons as well as to aggregates of baryons due to their non-zero baryon number. We called the baryonic gauge field particle a *planckton*. Its electromagnetic analogue is the photon.

Plancktons propagate in the Megaverse, both within universes, and exterior to universes. So the planckton field must be defined in 16-dimensional Megaverse coordinates. They will interact with baryons within a universe with Megaverse coordinates mapped to the curved coordinates of the universe. Since a planckton field in 16-dimensional conventional coordinates would lead to divergences we use 16-dimensional quantum coordinates:[89]

$$Y^i(y) = y^i + i\, Y_u^{\,i}(y)/M_u^{\,8} \tag{15.7}$$

with quantum coordinate derivatives defined by

$$\partial_i = \partial/\partial Y^i(y) = \partial/\partial(y^i - Y_u^{\,i}(y)/M_u^{\,8}) \tag{15.8}$$

to obtain a completely finite theory of planckton interactions with elementary particles and universe particles.

Plancktons and the $Y_u^{\,i}(y)$ field of quantum coordinates are the only fields in the space between universes in the Megaverse. Since the mass-energy and charge of universes is zero, Standard Model fields are zero in the space between universes.[90] It is reasonable to assume that the vacuum between universes does have fermion and boson seas for Standard Model particles.

We will propose new forces: Dark baryon force, lepton force and Dark lepton force corresponding to three other particle numbers in section 15.2 and in chapter 16. These four forces will have gravitational effects in the Megaverse.

15.1.3 Planckton Second Quantization

The second quantization of the free planckton field $B_u^{\,i}(y)$ is similar to the second quantization of the electromagnetic field, and also of the quantum part of the Megaverse quantum coordinates $Y_u^{\,i}(y)$. The purpose and role of these fields is quite different: the planckton field generates an interaction between baryons while the $Y_u^{\,i}(y)$ field serves as the

---

[87] The Megaverse is a 16-dimension complex space within which universes are embedded. We will discuss its features chapter 23. The analysis presented here would hold in our 4-dimensional space-time with trivial differences.

[88] Compared to gravity.

[89] See Blaha (2005a) for a discussion of this new method to eliminate infinities in quantum field theory calculations.

[90] The vacuum energy of the baryonic field and the $Y_u^{\,i}(y)$ fields being uniform throughout the Megaverse do not exert forces or cause gravitational effects except possibly through baryonic Casimir forces between universes.

quantum part of 4-dimensional (or 16-dimensional) quantum coordinates giving us a finite quantum field theory of The New Standard Model and gravitation as well as a finite Big Bang for our universe.

We begin by noting that Megaverse quantum coordinates are defined by eqns. 15.7 and 15.8 above. The lagrangian density terms for the free $B_u^{\,i}(Y(y))$ fields is

$$\mathscr{L}_{Bu} = -\tfrac{1}{4}\, F_{Bu}{}^{\mu\nu}(Y(y))F_{Bu\mu\nu}(Y(y)) \qquad (15.9)$$

with Y(y) given by eq. 15.7. The lagrangian is

$$L_{Bu} = \int d^{15}y\, \mathscr{L}_{Bu}(Y(y)) \qquad (15.10)$$

with

$$F_{Bu\mu\nu} = \partial B_{u\mu}(Y(y))/\partial Y^{\nu}(y) - \partial B_{u\nu}(Y(y))/\partial Y^{\mu}(y) \qquad (15.11)$$

where the values of μ and ν range from 1 to 16 in this section.

The equal time commutation relations, derived in the usual way, are:

$$[B_u{}^{\mu}(Y(\mathbf{y}, y^0)), B_u{}^{\nu}(Y(\mathbf{y}', y^0))] = [\pi_u{}^{\mu}(Y(\mathbf{y}, y^0)), \pi_u{}^{\nu}(Y(\mathbf{y}', y^0))] = 0 \qquad (15.12)$$

$$[\pi_{uj}(Y(\mathbf{y}, y^0)), B_{uk}(Y(\mathbf{y}', y^0))] = -i\,\delta^{15tr}{}_{jk}(Y(\mathbf{y},0) - Y(\mathbf{y}',0)) \qquad (15.13)$$

where

$$\pi_u{}^{k} = \partial L_u\,(B_u(Y(y)))/\partial B_{uk}'(Y(y)) \qquad (15.14)$$

$$\pi_u{}^{0} = 0 \qquad (15.15)$$

and

$$\delta^{tr}{}_{jk}(\mathbf{y} - \mathbf{y}') = \int d^{15}k\; e^{i\,\mathbf{k}\bullet(Y(\mathbf{y},0) - Y(\mathbf{y}',0))}\,(\delta_{jk} - k_jk_k/\mathbf{k}^2)/(2\pi)^{15} \qquad (15.16)$$

$$B_{uk}'(Y(y)) = \partial B_{uk}(Y(y))/\partial y^{16} \qquad (15.17)$$

for j, k = 1, 2, … , 15.

If we choose the Coulomb gauge for $B_{uk}(Y(y))$:

$$B_u{}^{16}(Y(y)) = 0$$

$$\partial B_u{}^{j}(Y(y))/\partial Y^{j}(y) = 0$$

for j = 1, 2, … , 15 then fourteen degrees of freedom (polarizations) are present in the vector potential.[91] The Fourier expansion of the vector potential $B_u{}^{i}(Y(y))$ is:

$$B_u{}^{i}(Y(y)) = \int d^{15}k\; N_{0B}(k) \sum_{\lambda=1}^{14} \varepsilon^{i}(k, \lambda)[a_B(k,\lambda) :e^{-ik\cdot Y(y)}: + a_B{}^{\dagger}(k,\lambda) :e^{ik\cdot Y(y)}:] \qquad (15.18)$$

[91] Note we use the Coulomb gauge for Y(y) also.

for i = 1, … , 15 where

$$N_{0B}(k) = [(2\pi)^{15} 2\omega_k]^{-½} \quad (15.19)$$

and (since the field is massless)

$$k^{16} = \omega_k = (\mathbf{k}^2)^{½} \quad (15.20)$$

where k^{16} is the energy, and where the $\varepsilon^i(k, \lambda)$ are the polarization unit vectors for $\lambda = 1, \ldots, 14$ and $k^\mu k_\mu = k^{16\,2} - \mathbf{k}^2 = 0$.

The commutation relations of the Fourier coefficient operators are:

$$[a_B(k,\lambda), a_B^\dagger(k',\lambda')] = \delta_{\lambda\lambda'}\delta^{15}(\mathbf{k} - \mathbf{k}') \quad (15.21)$$
$$[a_B^\dagger(k,\lambda), a_B^\dagger(k',\lambda')] = [a_B(k,\lambda), a_B(k',\lambda')] = 0 \quad (15.22)$$

and the polarization vectors satisfy

$$\sum_{\lambda=1}^{14} \varepsilon_i(k, \lambda)\varepsilon_j(k, \lambda) = (\delta_{ij} - k_i k_j/\mathbf{k}^2) \quad (15.23)$$

The B_u^μ Feynman propagator is

$$iD_F^{trTT}(y_1 - y_2)_{jk} = <0|T(B_{uj}(Y(y_1))B_{uk}(Y(y_2)))|0> \quad (15.24)$$

$$= -\, ig_{jk} \int \frac{d^{16}k\, e^{-ik.(y_1 - y_2)}\, R(\mathbf{k}, y_1 - y_2)}{(2\pi)^{16}\,(k^2 + i\varepsilon)} \quad (15.25)$$

where g_{jk} is the 16-dimensional Lorentz metric and where $R(\mathbf{k}, y_1 - y_2)$ is given by

$$R(\mathbf{k}, y_1 - y_2) = \exp[-\, k^i k^j \Delta_{Tij}(y_1 - y_2)/M_u^{16}] \quad (15.26)$$
$$= \exp\{-k^2[A(v) + B(v)\cos^2\theta] / [(2\pi)^{14} M_u^4 z^2]\}$$

where k^2 is the sum of the squares of the 15 spatial components with

$$z^\mu = y_1^\mu - y_2^\mu$$
$$z = |\mathbf{z}| = |\mathbf{y}_1 - \mathbf{y}_2|$$
$$k = |\mathbf{k}|$$
$$v = |z^0|/z$$
$$A(v) = (1 - v^2)^{-1} + .5v\ \ln[(v - 1)/(v + 1)]$$
$$B(v) = v^2(1 - v^2)^{-1} - 1.5v\ \ln[(v - 1)/(v + 1)]$$
$$\mathbf{k}\cdot\mathbf{z} = kz\cos\theta$$

and |**k**| denoting the length of a spatial 15-vector **k** while $|z^0|$ is the absolute value of $z^0 \equiv z^{16}$.

As eq. 15.26 indicates, the Gaussian damping factor R(k, z) for all large spatial momentum k^j is the same for both the positive and negative frequency parts of the (Two-Tier) B_u Feynman propagator. We are assuming the spatial momentum is real-valued in this discussion. It is also important to note that R(k, z) does not depend on $k^0 = k^{16}$ (in the B_u and Y_u Coulomb gauges) and thus the integration over k^0 proceeds in the usual way to produce time-ordered positive and negative frequency parts.

The Gaussian exponential factor in *all* spatial coordinates causes the Feynman propagator to be finite and, together with the Gaussian factor in universe particle propagators, causes all perturbation theory calculations, when interactions are introduced, to be finite as we have seen earlier in The Extended Standard Model.

For small momentum, much less than M_u, then $R(\mathbf{k}, y_1 - y_2) \rightarrow 1$ and the Feynman propagator is the "normal" propagator of conventional 16-dimensional quantum field theory. For large momentum the corresponding potential approaches r^{13} in contrast to the electromagnetic Coulomb potential r^{-1}. The B_u potential is highly non-singular at large energies.

15.1.4 Bary-Electric Fields and Bary-Magnetic Fields

As in electromagnetism there is an antisymmetric tensor of the second rank that appears in the free part of the baryonic field $F_{Bu\mu\nu}(y)$ lagrangian:[92]

$$\mathscr{L}_{Bu} = -\tfrac{1}{4} F_{Bu}{}^{ij}(y) F_{Buij}(y) \tag{15.27}$$

where

$$F_{Buij}(y) = \partial B_{ui}(y)/\partial y^j - \partial B_{uj}(y)/\partial y^i \tag{15.28}$$

and i, j = 1, 2, ... , 16. The 16th coordinate corresponds to the time coordinate. While the coordinates are complex in general we will treat the 15 spatial coordinates as real and the 16th coordinate as pure imaginary with the resulting invariant interval

$$ds^2 = dy_1{}^2 + dy_2{}^2 + \ldots + dy_{15}{}^2 - c^2 dy_{16}{}^2 \tag{15.29}$$

which is invariant under 16 dimensional Lorentz transformations. The coordinates can be transformed into complex-valued coordinates using the Reality group defined in Blaha (2014) and earlier books.

The tensor F_{Buij} is conveniently separated into a baryon electric part and a baryon magnetic part in a manner similar to the separation of the electromagnetic fields into electric

---

[92] Parts of the following appear in Blaha (2014a). They are somewhat modified since we are dealing with the classical, low energy, large distance baryonic field where the quantum coordinate fields Y(y) are well approximated by the classical (non-quantum) Megaverse coordinates y.

and magnetic fields. However the 15 spatial dimensions changes the forms of the baryon fields. In analogy to electromagnetism the baryon force is given by

$$f_i = F_{Buij}(y)J_B^j/c \qquad (15.30)$$

where J_B^j is the jth baryonic current.

The baryon "electric" field is

$$E_{Bui} = -F_{Bui0}(y)/c \qquad (15.31)$$

while the baryon "magnetic" field is

$$B_{Bui} = \varepsilon_{ijk}F_{Bu}{}^{jk}(y) \qquad (15.32)$$

where i, j, k = 1, 2, … , 15 and where ε_{ijk} is a totally anti-symmetric tensor with component values ±1. If $i < j < k$ then ε_{ijk} is +1. Even permutations of these three indices yield a value of +1 for the tensor components. Odd permutations of these three indices yield a value of –1. For example, $\varepsilon_{246} = +1$, $\varepsilon_{426} = -1$, $\varepsilon_{642} = -1$, $\varepsilon_{264} = -1$, $\varepsilon_{462} = +1$, $\varepsilon_{624} = +1$.

With these definitions of the $\mathbf{E_{Bu}}$ and $\mathbf{B_{Bu}}$ fields we can easily derive the 16-dimensional generalization of the *Lorentz force law* for a baryon of charge q and 15-velocity v_j:

$$F_i = qE_{Bui} + q\varepsilon_{ijk}v_jB_{Buk}/c \qquad (15.33)$$

for i = 1, 2, … , 15. One important difference from the 4-dimensional case is the forms of the $\mathbf{E_{Bu}}$ and $\mathbf{B_{Bu}}$ fields

$$E_{Bui} = -F_{Bui0}(y)/c = [-\partial B_{u0}(y)/\partial y^i - \partial B_{ui}(y)/\partial y^0] \qquad (15.34)$$

or, expressed as a 15-vector,

$$\mathbf{E_{Bu}} = [-\nabla_{15}\phi(y) - \mathbf{B_u}(y)]/c \qquad (15.35)$$

where ϕ is the baryonic Coulomb potential $B_{u16}(y)$, $\nabla 15$ is the 15-dimensional grad operator, and $B_u(y)$ is the baryonic 15-vector potential with the "dot" above it signifying a time (y_{16}) derivative.

The 15-dimensional baryon magnetic field has the form of eqn. 15.32. A specific illustrative case shows the baryon magnetic field exhibits more complexity than the 3-dimensional magnetic field of electromagnetism:

$$B_{Bu1} = \varepsilon_{1jk}F_{Bu}{}^{jk}(y)/c = [F_{Bu}{}^{23}(y) + F_{Bu}{}^{24}(y) + \ldots + F_{Bu}{}^{215}(y) + F_{Bu}{}^{34}(y) + F_{Bu}{}^{35}(y) + \ldots + F_{Bu}{}^{315}(y) + F_{Bu}{}^{45}(y) + \ldots + F_{Bu}{}^{14,15}(y)]/c \qquad (15.36)$$

Thus each component of the baryon magnetic field impacts on all fifteen spatial directions of the Megaverse. For this reason we use spinning rings, mass configurations and uniships to generate baryon magnetic field interactions to enable uniships to escape from our universe's three spatial dimensions. We consider this possibility in more detail in the following sections.

15.1.5 The Baryonic "Coulombic" Gauge Field

The baryonic gauge field has a "Coulombic" potential part $\phi(y)$, just as the electromagnetic field does. Consequently the total potential between two electromagnetically neutral masses of mass M_1 and M_2, and baryon numbers N_1 and N_2 is

$$V_{tot} = -GM_1M_2/r + (\beta^2/4\pi)\, N_1N_2/r \tag{15.37}$$

where G is the gravitational constant, and β is analogous to the electric charge e in the electromagnetic Coulomb potential. If both masses are composed of the same substance, and have the same mass, then we can set $M_1 = M_2 = M = Nm$ where m is the average mass of the baryons in the masses.[93] In addition we can set $N_1 = N_2 = N$. Then eq. 15.37 becomes

$$V = [-Gm^2 + (\beta^2/4\pi)]N^2/r \tag{15.38}$$

Note that the gravitational potential term is attractive, and the baryonic potential term is repulsive between baryons.

In considering eq. 15.1 we have approximated the baryonic potential with only our universe's spatial coordinates. In reality we should be using the spatial separation in all Megaverse coordinates. However since our universe is close to flat, the distance between two objects that are not too far apart is approximately the same in both coordinate systems. The baryonic potential in Megaverse coordinates is actually

$$\phi(y_1, y_2, \ldots, y_{15}) = (\beta^2/4\pi)N_1N_2/(y_1^2 + y_2^2 + \ldots + y_{15}^2)^{1/2} \tag{15.39}$$

15.1.6 The Baryonic Force on Baryonic Objects

The baryonic force on a moving baryon mass is given by the baryon Lorentz force for a baryon of baryon charge q and 15-velocity v_j:

$$F_i = qE_{Bui} + q\varepsilon_{ijk}v_jB_{Buk}/c \tag{15.40}$$

for i = 1, 2, … , 15. The 16-dimensional baryonic Coulomb potential is

$$V = N\phi(y_1, y_2, \ldots, y_{15}) = (\beta^2/4\pi)N/(y_1^2 + y_2^2 + \ldots + y_{15}^2)^{1/2} \tag{15.41}$$

---

[93] We neglect lepton masses since they are negligible relative to the baryon masses.

where N is the baryon number of the baryon mass. The baryon Coulomb force is

$$F_i = N\nabla_{15i}\phi(y) \tag{15.42}$$

where ∇_{15i} is the ith component of the 15-dimensional grad operator $\boldsymbol{\nabla}_{15}$.

15.2 Lepton Number, Dark Baryon Number, and Dark Lepton Number Conservation Laws

In the previous section we raised the possibility of an ultra-weak Baryonic force and an associated Baryon number conservation law. This conservation law has been repeatedly tested and found to be satisfied to great accuracy. One possible cause for concern is the Adler-Bell-Jackiw fermion triangle anomaly, which follows from a three fermion loop that diverges in conventional quantum field theory. This type of anomaly raises the possibility of baryon number non-conservation. In Two-Tier Quantum Field Theory the triangle graph is convergent – no infinities – and anomalies are not present. Thus our Two-Tier implementation of the Extended Standard Model is anomaly-free and the issue of baryon number non-conservation disappears. (See chapter 19 and Blaha (2005a) for more detailed discussions.)

We will assume Baryon number conservation holds.

15.2.1 Other Conserved Particle Numbers

Another conserved number is Lepton Number, denoted L. Again, repeated attempts to find lepton number violation have failed. On that basis *we will assume lepton number conservation.*

If Baryon Number and Lepton Number are both conserved quantities then any linear combination of them is also conserved. Therefore

$$B' = aB + bL \tag{15.43}$$

is also conserved.

If we consider the Dark Matter sector of the Extended Standard Model it is reasonable to assume that *Dark Baryon Number B_D and Dark Lepton Number L_D are conserved* also although there is no experimental evidence available as yet to confirm (or deny) these assumptions.

Thus we have four conserved particle Numbers. Linear combinations of these numbers are also conserved:

$$\begin{aligned} B' &= aB + bL + cB_D + dL_D \\ L' &= eB + fL + gB_D + hL_D \\ B_D' &= iB + jL + kB_D + lL_D \\ L_D' &= mB + nL + oB_D + pL_D \end{aligned} \tag{15.44}$$

or

$$N' = AN \tag{15.45}$$

where N and N' are 4-vectors composed of particle numbers and A is a 4×4 matrix. The number of fermion particle types, 4, is determined by the two families of normal fermions, leptons and quarks, and the number of families of Dark fermions, which is ultimately determined by the Reality group study of fermion types in chapter 5.

The constants appearing in linear equations of the form of eq. 15.45 seem arbitrary. However if we want the new 'primed' set of conserved Numbers to be an independent set of numbers then the determinant of the constants must be non-zero. Thus the matrix, A, is invertible.

The set of 4×4 matrices of the type of eq. 15.44 form an U(4) group[94] *if we wish to perform these transformations within lagrangians of the type of the Extended Standard Model. The choice of U(4) rather than SU(4) is required Since there are four independent particle Numbers and U(4) has four diagonal matrices in its algebra while SU(4) only has three diagonal matrices. U(4) preserves the independence of the four independent particle Numbers.*

15.3 U(4) Number Symmetry

At this point we note the observations of Yang and Mills that Numbers can be local and generalize the U(4) Number symmetry to a Yang-Mills symmetry. The transformations then become functions of position A(X) where X represents the space-time coordinates in Two-Tier Quantum Field Theory.

The U(4) rotations of the four Numbers changes the interpretation of the number operators applied to particle fields. Thus we have a symmetry operation induced on particle fields, which in the absence of symmetry brealing terms, becomes a symmetry of the lagrangian.

To implement this symmetry in the particles' lagrangian, all covariant derivatives must acquire another interaction term with 16 U(4) fields corresponding to the 16 generators of U(4). In addition we must add another index to each fermion field specifying its generation. Lastly a set of initially massless gauge field dynamic terms must be added to the Extended Standard Model lagrangian.

The initially massless U(4) gauge transformation symmetry is broken by the Higgs Mechanism with the gauge fields acquiring masses (with two exceptions) and the fermions of the four generations acquiring masses which are generation dependent. The symmetry breaking is described in the next chapter.

16. Fermion Generations and Broken U(4) Generation Group

In sections 15.2 and 15.3 we showed that a local U(4) symmetry based on conserved fermion particle numbers existed in Nature and added a new assumption of a U(4) Generation

[94] If we wish to further limit the values of the 'primed' Numbers to integers assuming the unprimed Numbers are integers then the group of the transformation is the set of permutations of four entities – the Symmetric group S_4. However the 'primed' numbers can be integer or not. There is no apparent physical principle requiring integer Numbers. Quarks are usually assigned Baryon Number 1/3. Also Numbers are not necessarily positive valued.

group to our construction of the Extended Standard Model. U(4) has 16 generators, which we denote G_i for i = 1, 2, … , 16. Its fundamental representation has 4×4 matrices. When we introduce this U(4) symmetry directly into the one generation Extended Standard Model each fermion acquires a new index and becomes a four generation set of fermions. Symmetry breaking via a Higgs Mechanism for the U(4) gauge fields gives a different mass to each of the members of each set. The Higgs particle lagrangian for U(4) breaking will be described in section 16.3.

16.1 Four Generation Extended Standard Model

In chapter 14 we derived the form of a one generation Extended Standard Model that included the known parts of the Standard Model (excepting the Higgs sector) and an SU(2)⊗U(1) part for Dark Matter. Dark Matter was linked to normal matter with a simple scalar gauge field. [Now it is replaced by a new U(4) Layer group interaction shown later.]

In this section we generalize to the four generation Extended Standard Model that results.[95] Covariant derivatives acquire another interaction term with 16 U(4) fields U_i^{μ}. In addition we add another index to each fermion field specifying its generation. Lastly a set of initially massless gauge field dynamic terms is added to the Extended Standard Model lagrangian to specify U(4) gauge field evolution.

16.1.1 Two-Tier Lepton Sector

We begin with the definition of a quadruplet of leptons – a pair of doublets, one normal and one Dark, instead of a single doublet. We define left and right lepton quadruplets with[96]

$$\Psi_{L,Ra}(X) = \begin{bmatrix} \psi_{DL,Ra}(X) \\ \psi_{NL,Ra}(X) \end{bmatrix} \tag{16.1}$$

where a is a generation index ranging from 1 to 4, where $\psi_{NL,R}(X)$ is a "normal" ElectroWeak-like lepton doublet consisting of a normal electron-like fermion and a normal neutrino-like fermion, and where $\psi_{DL,R}(X)$ is a Dark ElectroWeak-like lepton doublet consisting of a Dark electron-like fermion and a Dark neutrino-like fermion.

We define covariant derivative terms, which we express in matrix form as

[95] It is based on the three principles resulting from applying Ockham's Razor ("The simplest choice is often the best."): 1) The only connecting interaction is a weak interaction, 2) The form of ElectroWeak theory remains unchanged, and 3) Dark Matter parallels normal matter in its general characteristics: four generations, SU(3) singlets, an SU(2)⊗U(1) symmetry analogous to ElectroWeak symmetry, SU(2)⊗U(1) Dark lepton and Dark quark doublets.

[96] The X's are Two-Tier coordinates.

$$D_{L,R}(X) = \begin{bmatrix} \gamma^\mu D_{DL,R\mu} & 0 \\ 0 & \gamma^\mu D_{NL,R\mu} \end{bmatrix} \tag{16.2}$$

where the normal matter left-handed covariant derivative is

$$D_{NL\mu} = \partial/\partial X^\mu - \tfrac{1}{2}ig\boldsymbol{\sigma}\cdot\mathbf{W}_\mu + \tfrac{1}{2}ig'B_\mu - \tfrac{1}{2}ig_G\mathbf{G}\cdot\mathbf{U}_\mu \tag{16.3}$$

where g_G is an ultra-weak generation coupling constant, and where $\mathbf{G}\cdot\mathbf{U}_\mu$ is the sum of the inner product of 16 U(4) generators G_i and gauge fields $U_{\mu i}(X)$. The Dark matter left-handed covariant derivative is

$$D_{DL\mu} = \partial/\partial X^\mu - \tfrac{1}{2}ig_D\boldsymbol{\sigma}\cdot\mathbf{W}'_\mu + \tfrac{1}{2}ig_D'B'_\mu + \tfrac{1}{2}ig_D''B_\mu - \tfrac{1}{2}ig_G\mathbf{G}\cdot\mathbf{U}_\mu \tag{16.4}$$

with $\boldsymbol{\sigma}$ a vector composed of the Pauli matrices. The right-handed covariant derivatives have a simpler form. The normal matter right-handed covariant derivative is

$$D_{NR\mu} = \partial/\partial X^\mu + \tfrac{1}{2}ig'B_\mu - \tfrac{1}{2}ig_G\mathbf{G}\cdot\mathbf{U}_\mu \tag{16.5}$$

and the Dark matter right-handed covariant derivative is

$$D_{DR\mu} = \partial/\partial X^\mu + \tfrac{1}{2}ig_D'B'_\mu + \tfrac{1}{2}ig_D''B_\mu - \tfrac{1}{2}ig_G\mathbf{G}\cdot\mathbf{U}_\mu \tag{16.6}$$

The normal and Dark electroweak fields above are functions of Two-Tier coordinates X. The Faddeev-Popov mechanism operative for these types of fields is described in appendix 19-A of Blaha (2011c) and in chapter 12.

16.1.2 Quark Sector

In the *quark* sector we define left and right quark quadruplets with

$$\Psi_{qL,Ra}(X_c) = \begin{bmatrix} \psi_{DqL,Ra}(X_c) \\ \psi_{NqL,Ra}(X_c) \end{bmatrix} \tag{16.7}$$

where $\psi_{NqL,Ra}(X_c)$ is a "normal" ElectroWeak-like quark doublet consisting of an SU(3) color up-quark and a color SU(3) down-quark, and where $\psi_{DqL,Ra}(X_c)$ is a Dark ElectroWeak-like quark doublet consisting of an SU(3) singlet Dark up-quark of unit Dark charge and an SU(3) singlet Dark down-quark of zero Dark charge in the ath generation.

The covariant derivative terms are contained in $D_q(X_c)$ which we express in matrix form as

$$D_{qL,R}(X_c) = \begin{bmatrix} \gamma^\mu D_{qDL,R\mu}(X_c) & 0 \\ 0 & \gamma^\mu D_{qNL,R\mu}(X_c) \end{bmatrix} \tag{16.8}$$

where the normal quark matter left-handed covariant derivative is

$$D_{qNL\mu} = \partial/\partial X_c{}^\mu - \tfrac{1}{2} ig\boldsymbol{\sigma}\cdot\mathbf{W}_\mu - ig'B_\mu/6 - \tfrac{1}{2} ig_G\mathbf{G}\cdot\mathbf{U}_\mu + ig_C\tau\cdot A_{C\mu} \tag{16.9}$$

and where the Dark quark left-handed covariant derivative is

$$D_{qDL\mu} = \partial/\partial X_c{}^\mu - \tfrac{1}{2} ig_D\boldsymbol{\sigma}\cdot\mathbf{W}'_\mu + \tfrac{1}{2} ig_D'B'_\mu + \tfrac{1}{2} ig_D''B_\mu - \tfrac{1}{2} ig_G\mathbf{G}\cdot\mathbf{U}_\mu \tag{16.10}$$

since Dark quarks are SU(3) singlets with unit or zero Dark charge. The right-handed quark covariant derivatives have a simpler form. The normal quark right-handed covariant derivative is

$$D_{qNR\mu} = \partial/\partial X_c{}^\mu + \tfrac{1}{2} ig'B_\mu/3 - \tfrac{1}{2} ig_G\mathbf{G}\cdot\mathbf{U}_\mu + ig_C\tau\cdot A_{C\mu} \tag{16.11}$$

and the Dark quark right-handed covariant derivative is

$$D_{qDR\mu} = \partial/\partial X_c{}^\mu + \tfrac{1}{2} ig_D'B'_\mu + \tfrac{1}{2} ig_D''B_\mu - \tfrac{1}{2} ig_G\mathbf{G}\cdot\mathbf{U}_\mu \tag{16.12}$$

The normal and Dark gauge fields are functions of Two-Tier coordinates, $X_c = (X_{r\mu}(y_r), X_{i\mu}(y_i))$, of eqs. 14.11 and 14.12. The Faddeev-Popov mechanism is operative for gauge boson fields and is described in appendix 19-A of Blaha (2011c).[97] The *complexon* quark Extended Standard Model ElectroWeak Sector covariant derivatives in quadruplet matrix form are

$$D_{qL,R}(X_c) = \begin{bmatrix} \gamma^\mu D_{qDL,R\mu} & 0 \\ 0 & \gamma^\mu D_{qNL,R\mu} \end{bmatrix} \tag{16.13}$$

The remaining parts of the complexon Standard Model are described in chapter 23 of Blaha (2011) and summarized below. The addition of singlet Dark quark Higgs terms is also required.

[97] Those who might be concerned about the propagator term $\langle W_i(X), W_j(X_c)\rangle$ and similar propagators where one field is a function of X and the other field is a function of X_c should note that such terms are to very good approximation equal to $\langle W_i(X), W_j(X)\rangle$ for energies much less than M_c. (These energies could be as large as the Planck energy.)

The lagrangian density and action is

$$\mathcal{L}_{CSM} = \Psi_L^{\dagger}\gamma^0 i\gamma^\mu D_{L\mu}\Psi_L - \Psi_R^{\dagger}\gamma^0 i\gamma^\mu D_{R\mu}\Psi_{3R} + \Psi_{CL}^{\dagger}\gamma^0 i\gamma^\mu \mathcal{D}_{qL\mu}\Psi_{CL} + \Psi_{CR}^{\dagger}\gamma^0 i\gamma^\mu \mathcal{D}_{qR\mu}\Psi_{CR} - - \mathcal{L}_{BareMasses} + \mathcal{L}_{Gauge} + \mathcal{L}_{Mass} + \mathcal{L}_{Ufields} \quad (16.14)$$

where there is an implicit sum over generations. $\mathcal{L}_{BareMasses}$ contains the fermion bare mass terms. Also,

$$\mathcal{L}_{Gauge} = \mathcal{L}_{GaugeEW} + \mathcal{L}_{GaugeC} + \mathcal{L}_{GaugeEWD} \quad (16.15)$$

with

$$\mathcal{L}_{GaugeEW} = -\tfrac{1}{4} F_W{}^{a\mu\nu} F_W{}^a{}_{\mu\nu} - \tfrac{1}{4} F_B{}^{\mu\nu} F_{B\mu\nu} + \mathcal{L}_{EW}{}^{ghost} \quad (16.16)$$

$$\mathcal{L}_{GaugeEWD} = -\tfrac{1}{4} F'_W{}^{a\mu\nu} F'_W{}^a{}_{\mu\nu} - \tfrac{1}{4} F_{B'}{}^{\mu\nu} F_{B'\mu\nu} + \mathcal{L}_{W'}{}^{ghost} \quad (16.17)$$

and

$$\mathcal{L}_{GaugeC} = \mathcal{L}_{CCG} + \mathcal{L}_C{}^{ghost} + \mathcal{L}_{CC}{}^{ghost} \quad (16.18)$$

$$\mathcal{L}_{Ufields} = -\tfrac{1}{4} F_U{}^{a\mu\nu} F_{U\mu\nu\nu} + \mathcal{L}_U{}^{ghost} + \mathcal{L}_U{}^{UHiggs} \quad (16.19)$$

where $\mathcal{L}_U{}^{UHiggs}$ is discussed in section 16.4. The ElectroWeak gauge bosons $W_\mu{}^a$, B_μ and B'_μ field tensors are:

$$F_W{}^a{}_{\mu\nu} = \partial W^a{}_\mu/\partial X^\nu - \partial W^a{}_\nu/\partial X^\mu + g_2 f^{abc} W^b{}_\mu W^c{}_\nu \quad (16.20)$$

$$F_{B\mu\nu} = \partial B_\mu/\partial X^\nu - \partial B_\nu/\partial X^\mu \quad (16.21)$$

and the Dark ElectroWeak gauge bosons $W'_\mu{}^a$ and B'_μ field tensors are:

$$F_{B'\mu\nu} = \partial B'_\mu/\partial X^\nu - \partial B'_\nu/\partial X^\mu$$

$$F'_W{}^a{}_{\mu\nu} = \partial W'^a{}_\mu/\partial X^\nu - \partial W'^a{}_\nu/\partial X^\mu + g_2 f^{abc} W'^b{}_\mu W'^c{}_\nu \quad (16.22)$$

The U fields' tensor is:

$$F_U{}^a{}_{\mu\nu} = \partial U^a{}_\mu/\partial X^\nu - \partial U^a{}_\nu/\partial X^\mu + g_G f_4{}^{abc} U^b{}_\mu U^c{}_\nu \quad (16.23)$$

where $f_4{}^{abc}$ are the U(4) algebra structure constants.

$\mathcal{L}_{EW}{}^{ghost}$ contains the Faddeev-Popov ghost terms for the ElectroWeak $W_\mu{}^a$ gauge bosons. The complexon color gluon lagrangian $\mathcal{L}_{CCG}$ is defined by

$$\mathcal{L}_{CCG} = -\tfrac{1}{4}\, F_{CC}{}^{a\mu\nu}(X) F_{CC}{}^{a}{}_{\mu\nu}(X) \tag{16.24}$$

where

$$F_{CC}{}^{a}{}_{\mu\nu} = \partial/\partial X_c{}^{\nu}\, A_C{}^{a}{}_{\mu} - \partial/\partial X_c{}^{\mu}\, A_C{}^{a}{}_{\nu} + g f_{su(3)}{}^{abc} A_C{}^{b}{}_{\mu} A_C{}^{c}{}_{\nu} \tag{16.25}$$

where $A_C{}^{a}{}_{\nu}$ is the color gluon gauge field, g is the color coupling constant, and the $f_{su(3)}{}^{abc}$ are the SU(3) structure constants.

In addition $\mathcal{L}_C{}^{ghost}$ is the color SU(3) Faddeev-Popov ghost terms defined in appendix 19-A of Blaha (2011c) for the complexon Lorentz gauge and $\mathcal{L}_{CC}{}^{ghost}$ is the complexon color SU(3) constraint ghost terms defined through the Faddeev-Popov mechanism. The mass sector $\mathcal{L}_{Mass}$ is presumably based on the Higgs Mechanism, which creates the fermion and ElectroWeak vector boson masses, and generation mixing.

The lagrangian is supplemented with the following condition on all complexon fields $\Phi_{\dots}$:[98]

$$\nabla_r \cdot \nabla_i \Phi\ldots = 0 \tag{16.26}$$

Non-complexon fields Ω... in the left-handed formulation under consideration satisfy the subsidiary condition:

$$[\nabla_r \cdot \nabla_i - (\nabla_r^2 \nabla_i^2)^{1/2}]\Omega\ldots = 0 \tag{16.27}$$

which guarantees a complexon's real 3-momentum is parallel to its imaginary 3-momentum.

16.2 Generation U(4) Gauge Symmetry Breaking and Long Range Forces

In chapter 15 we showed that there was good experimental evidence for a conserved Baryon Number B and we proceeded to develop a simple U(1) gauge theory that would imply Baryon Number conservation in a manner analogous to QED's implying electric charge conservation. In section 16.1 we used a new symmetry group local U(4) to generalize the one generation Extended Standard Model to a four generation Extended Standard Model based on four conserved particle numbers: B, L, B_D, and L_D.[99]

We now assume in our construction that the four generation Extended Standard Model has a local U(4) symmetry that is broken by mass terms generated by the Higgs Mechanism.

Further, we will assume that the Higgs breakdown yields two massless (long-range) fields, which we associate with Baryon Number B and Dark Baryon Number B_D. The remaining fields acquire masses and generate short-range forces.

We use the following U(4) diagonal matrices:

$$G_1 = \text{diag}(1, 1, 1, 1) \tag{16.28}$$

[98] These conditions implement the orthogonality of the real and imaginary parts of complexon 3-momentum.

[99] Charge, although a conserved number, is a part of the ElectroWeak sector, account of which has already been taken.

$$G_2 = \mathrm{diag}(0, 1, 0, 0)$$
$$G_3 = \mathrm{diag}(0, 0, 1, 0)$$
$$G_4 = \mathrm{diag}(0, 0, 0, 1)$$

The U(4) algebra has 16 hermitean matrices that satisfy

$$G_i^\dagger = G_i \tag{16.29}$$

The particle numbers can be expressed in terms of the diagonal generators as

$$\begin{aligned} B &= G_1 - G_2 - G_3 - G_4 \\ B_D &= G_2 \\ L &= G_3 \\ L_D &= G_4 \end{aligned} \tag{16.30}$$

The covariant derivatives have the general form:

$$D_{\ldots\mu} = \partial/\partial X^\mu + \ldots - \tfrac{1}{2} i g_G \mathbf{G}\cdot\mathbf{U}_\mu \tag{16.31}$$

where the ellipsis's indicates the other details of the particular covariant derivative. We now wish to express the four gauge fields $U_i(X)$ for i = 1, 2, 3, 4 corresponding to the diagonal generators in terms of the fields of the four particle number gauge fields: B_μ, L_μ, $B_{D\mu}$, and $L_{D\mu}$.

$$U_{i\mu} = A_{ik} N_{k\mu} \tag{16.32}$$

where A_{ik} are the elements of a matrix of constants and

$$N_\mu = \begin{bmatrix} B_\mu(X) \\ L_\mu(X) \\ B_{D\mu}(X) \\ L_{D\mu}(X) \end{bmatrix} \tag{16.33}$$

is a column vector consisting of the gauge fields corresponding to each of the conserved particle numbers.

The matrix A must have non-zero determinant so that eq. 16.32 can be inverted to express the particle number fields in terms of the four $U_i(X)$ gauge fields:

$$N_\mu = A^{-1} U_\mu \tag{16.34}$$

resulting in

$$B_\mu(X) = U_{1\mu} \tag{16.35}$$
$$L_\mu(X) = U_{1\mu} + U_{2\mu}$$

$$B_{D\mu}(X) = U_{1\mu} + U_{3\mu}$$
$$L_{D\mu}(X) = U_{1\mu} + U_{4\mu}$$

Then

$$D_{\ldots\mu} = \partial/\partial X^{\mu} + \ldots - \tfrac{1}{2} i g_G[\sum_{i=5}^{16} \mathbf{G}_i \mathbf{U}_{i\mu} + BB_{\mu}(X) + LL_{\mu}(X) + B_D B_{D\mu}(X) + L_D L_{D\mu}(X)] \quad (16.36)$$

where the particle numbers, which are analogous to the charges Q and Q' in ElectroWeak theory, are B, L, B_D, and L_D. They are expressed in terms of U(4) generators by eqs. 16.30.

16.3 Higgs Mass Mechanism for Generation U(4) Gauge Fields

We now require that there are two massless fields: one coupled to Baryon number and one coupled to Dark Baryon number. The Dark sector is assumed to be analogous to the normal particle sector in this respect. Most of the fourteen remaining fields acquire masses and longitudinal components. These fields become short-range, ultra-weak generation forces. The masses they acquire through the Higgs Mechanism are presumably very large, as these gauge particles have not been found experimentally.[100]

We assume that a scalar Higgs field exists, which is a U(4) vector with four components corresponding to the fermion generations.[101] It is an SU(2)⊗U(1)⊗SU(3) ElectroWeak scalar. Its lagrangian density is

$$\mathcal{L}_U^{\text{UHiggs}} = (\partial\eta^{\dagger}/\partial X^{\mu})(\partial\eta/\partial X^{\mu}) - \lambda(\eta^{\dagger}\eta - \rho^2)^2 + \mathcal{L}_U^{\text{UHiggs}}{}_{\text{FermionMasses}}$$

where $\mathcal{L}_U^{\text{UHiggs}}{}_{\text{FermionMasses}}$ are the fermion masses produced by the U Higgs Mechanism and where we choose a unitary gauge in which the vector η is

$$\eta = \begin{bmatrix} 0 \\ \rho_1 \\ 0 \\ \rho_2 \end{bmatrix} \quad (16.37)$$

where ρ_1 and ρ_2 are Higgs fields with vacuum expectation values. Then the covariant derivative of η is

[100] Section 16.4 discusses this topic in more detail.

[101] We use the conventional Higgs particle formalism here because of its familiarity. Based on the preceding chapters the pseudoquantum formalism should be used where the vector $\eta = \varphi_1 + \varphi_2$ in an appropriately modified lagrangian. The non-zero vacuum expectation values of the φ_1 components are ρ_1 and ρ_2.

$$D_{\ldots\mu}\eta = \{\partial/\partial X^{\mu} + \ldots - \tfrac{1}{2} i g_G[\Sigma \mathbf{G}_i \mathbf{U}_{i\mu} + B B_{\mu}(X) + L L_{\mu}(X) + B_D B_{D\mu}(X) + L_D L_{D\mu}(X)]\} \begin{bmatrix} 0 \\ \rho_1 \\ 0 \\ \rho_2 \end{bmatrix} \tag{16.38}$$

The sum over i is from 5 through 16, and $[\mathbf{G}_i]_{jk}$ is the jk[th] element of $\mathbf{G}_i$. Then

$$D_{\ldots\mu}\eta = \begin{bmatrix} -\tfrac{1}{2} i g_G\{\rho_1\Sigma[\mathbf{G}_i]_{12}\mathbf{U}_{i\mu} + \rho_2\Sigma[\mathbf{G}_i]_{14}\mathbf{U}_{i\mu}\} \\ \partial\rho_1/\partial X^{\mu} - \tfrac{1}{2} i g_G \rho_1 L_{\mu} - \tfrac{1}{2} i g_G\{\rho_1\Sigma[\mathbf{G}_i]_{22}\mathbf{U}_{i\mu} + \rho_2\Sigma[\mathbf{G}_i]_{24}\mathbf{U}_{i\mu}\} \\ -\tfrac{1}{2} i g_G\{\rho_1\Sigma[\mathbf{G}_i]_{32}\mathbf{U}_{i\mu} + \rho_2\Sigma[\mathbf{G}_i]_{34}\mathbf{U}_{i\mu}\} \\ \partial\rho_2/\partial X^{\mu} - \tfrac{1}{2} i g_G \rho_2 L_{D\mu} - \tfrac{1}{2} i g_G\{\rho_1\Sigma[\mathbf{G}_i]_{42}\mathbf{U}_{i\mu} + \rho_2\Sigma[\mathbf{G}_i]_{44}\mathbf{U}_{i\mu}\} \end{bmatrix} \tag{16.39}$$

$$= \begin{bmatrix} -\tfrac{1}{2} i g_G\Sigma\{\rho_1[\mathbf{G}_i]_{12} + \rho_2[\mathbf{G}_i]_{14}\}\mathbf{U}_{i\mu} \\ \partial\rho_1/\partial X^{\mu} - \tfrac{1}{2} i g_G \rho_1 L_{\mu} - \tfrac{1}{2} i g_G \rho_2\Sigma[\mathbf{G}_i]_{24}\mathbf{U}_{i\mu} \\ -\tfrac{1}{2} i g_G\Sigma\{\rho_1[\mathbf{G}_i]_{32} + \rho_2[\mathbf{G}_i]_{34}\}\mathbf{U}_{i\mu} \\ \partial\rho_2/\partial X^{\mu} - \tfrac{1}{2} i g_G \rho_2 L_{D\mu} - \tfrac{1}{2} i g_G \rho_1\Sigma[\mathbf{G}_i]_{42}\mathbf{U}_{i\mu} \end{bmatrix} \tag{16.40}$$

since the generators $\mathbf{G}_i$ have zeroes along their diagonals for i = 5, … , 16.

From eq. 16.39 we find the corresponding Higgs field kinetic terms in the lagrangian are

$$(D_{\ldots\mu}\eta)^{\dagger} D^{\mu}_{\ldots}\eta = \partial\rho_1/\partial X^{\mu}\,\partial\rho_1/\partial X_{\mu} + \partial\rho_2/\partial X^{\mu}\,\partial\rho_2/\partial X_{\mu} + g_G^2\rho_1^2 L_{\mu} L^{\mu}/4 + g_G^2\rho_2^2 L_{D\mu} L_D^{\mu}/4 + \ldots \tag{16.41}$$

Note there are differing mass squared terms for the Lepton ($g_G^2\rho_1^2/4$) and Dark Lepton ($g_G^2\rho_2^2/4$) gauge fields making them short range fields with the likelihood of very large masses much beyond ElectroWeak gauge field masses, and with an ultra weak coupling constant g_G as suggested by the "experimental" coupling for the Baryonic force given in eq. 15.6.

The Baryonic and Dark Baryonic gauge fields are massless and thus long range although their coupling constant appears to be ultra weak – much below the gravitational coupling constant G.

We now turn to calculating the remaining terms in eq. 16.41 that determine the masses of the remaining 14 gauge fields. We begin by assigning matrix elements for the remaining hermitean U(4) generators:

$$
\begin{aligned}
[G_5]_{ik} &= \delta_{i1}\delta_{k2} + \delta_{i2}\delta_{k1} \\
[G_6]_{ik} &= -i\delta_{i1}\delta_{k2} + i\delta_{i2}\delta_{k1} \\
[G_7]_{ik} &= \delta_{i1}\delta_{k3} + \delta_{i3}\delta_{k1} \\
[G_8]_{ik} &= -i\delta_{i1}\delta_{k3} + i\delta_{i3}\delta_{k1} \\
[G_9]_{ik} &= \delta_{i1}\delta_{k4} + \delta_{i4}\delta_{k1} \\
[G_{10}]_{ik} &= -i\delta_{i1}\delta_{k4} + i\delta_{i4}\delta_{k1} \\
[G_{11}]_{ik} &= \delta_{i2}\delta_{k3} + \delta_{i3}\delta_{k2} \\
[G_{12}]_{ik} &= -i\delta_{i2}\delta_{k3} + i\delta_{i3}\delta_{k2} \\
[G_{13}]_{ik} &= \delta_{i2}\delta_{k4} + \delta_{i4}\delta_{k2} \\
[G_{14}]_{ik} &= -i\delta_{i2}\delta_{k4} + i\delta_{i4}\delta_{k2} \\
[G_{15}]_{ik} &= \delta_{i3}\delta_{k4} + \delta_{i4}\delta_{k3} \\
[G_{16}]_{ik} &= -i\delta_{i3}\delta_{k4} + i\delta_{i4}\delta_{k3}
\end{aligned}
\qquad (16.42)
$$

Then completing eq. 16.41 using eq. 16.40 we find

$$
\begin{aligned}
(D_{\ldots\mu}\eta)^\dagger D_{\ldots}^{\ \mu}\eta = {} & \partial\rho_1/\partial X^\mu \partial\rho_1/\partial X_\mu + \partial\rho_2/\partial X^\mu\, \partial\rho_2/\partial X_\mu + g_G^{\,2}\rho_1^{\,2}L_\mu L^\mu/4 + g_G^{\,2}\rho_2^{\,2}L_{D\mu} L_D^{\ \mu}/4 + \\
& + (g_G/2)^2\rho_1^{\,2}(U_5^{\,2} + U_6^{\,2}) + (g_G/2)^2\rho_2^{\,2}(U_9^{\,2} + U_{10}^{\,2}) + (g_G/2)^2\rho_1^{\,2}(U_{11}^{\,2} + U_{12}^{\,2}) + \\
& + (g_G/2)^2(\rho_1^{\,2} + \rho_2^{\,2})(U_{13}^{\,2} + U_{14}^{\,2}) + + (g_G/2)^2\rho_2^{\,2}(U_{15}^{\,2} + U_{16}^{\,2}) \qquad (16.43)
\end{aligned}
$$

up to total divergences, which generate surface terms which we discard, and assuming that all fields satisfy the gauge condition

$$\partial U_i^{\ \mu}/\partial X^\mu = 0 \qquad (16.44)$$

Note that there are no mass terms for $U_7(X)$ and $U_8(X)$ as well as $B_\mu(X)$ and $B_{D\mu}(X)$ due to our choice of unitary gauge eq. 16.37. Consequently there are four massless long range fields and 12 gauge fields that acquire masses of three different values: $(g_G/2)\rho_{10}$, $(g_G/2)\rho_{20}$, and $(g_G/2)(\rho_{10}^{\,2} + \rho_{20}^{\,2})^{½}$ where ρ_{10} and ρ_{20} ar the vacuum expectation values of ρ_1 and ρ_2 respectively.[102] The fields $U_7(X)$ and $U_8(X)$ are not "diagonal" and thus appear in the fermion sector as terms connecting fermions in different generations within the four species of normal fermions and within the four species of Dark fermions.[103] Therefore they do not change the values of any of the four types of particle numbers.

Based on the estimate of eq. 15.6 the ultra weak value of the coupling constant is

$$g_G = (4\pi\alpha_B)^{½} \approx 1.218\ (Gm_H^{\,2})^{½} \qquad (16.45)$$

The ultra-weak value of the coupling constant implies that the baryonic force with gauge field $B_\mu(X)$, which is now part of a quadruplet of fields, is a massless, long range field

[102] The Higgs fields ρ_i in our pseudoquantum formulation are $\rho_i = \varphi_{1i}(x) + \varphi_{2i}(x)$ with ρ_{i0} being the vacuum expectation of φ_{1i} as described earlier.

[103] Neutral lepton, charged lepton, up-type quark and down-type quark plus the four corresponding Dark species.

that corresponds to that of chapter 15 with the exception that chapter 15 looks ahead to later chapters where we discuss a 16-dimensional space that we call the *Megaverse* in which our universe resides where the baryonic force and force associated with the Dark Baryon force exist beyond our universe and act with other possible "island" universes. (The leptonic and Dark leptonic forces are short range and thus do not extend beyond our universe.)

The two non-diagonal long-range forces, being between different generations of a species and having an ultra-weak coupling, are not of great consequence because of the short lifetime of the higher generations of a species. Therefore, despite their long range, they have only the "shortest" time to exert an inter-generation force before a higher generation particle decays.

Since we expect the other massive fields to have very large masses (and thus very large Higgs field vacuum expectation values) and ultra-weak coupling they are not likely to be experimentally found for the foreseeable future.

16.4 Impact of this Generation U(4) Higgs Mechanism on Fermion Generation Masses

The fermion masses of the charged lepton, and the up-type quark, and down-type quark species' generations all show a rapid increase of mass with the generation. For example the u quark mass is a few MeV while the t quark (third generation) has a mass of about 170 GeV/c. The ratio of these masses is about 170,000. While one can account for this great difference by the judicious choices of Higgs' parameter values, when one considers the generational group and its associated numerical quantities: ultra-weak coupling, very large U particle masses – perhaps of the order of hundreds or thousands of GeV/c, and the corresponding very large Higgs particle vacuum expectation values in the U gauge field sector[104] then the differences in fermion masses within a species become more understandable and natural from a Leibniz Principle perspective. [The Layer group interactions add further mass terms, which also may be partly responsible for the large differences in mass between the generations of charged species. We discuss this possibility in more detail later.]

Thus the popular view that the ElectroWeak gauge field symmetry breaking occurs solely via ElectroWeak Higgs fields is not part of our Extended Standard Model unless the U(4) sector is removed. In our model there are two sets of contributions to fermion symmetry breaking: ElectroWeak Higgs particles symmetry breaking, and Generation group U(4) Higgs particles symmetry breaking [expended later to include a Layer group sector]. The Generation group causes each species to break into four generations.

[*The U(4) Generation group adds 12 more generators (and thus interaction terms) to the 4 ElectroWeak generators. The four generations requires a 4×4 matrix representation which we take to consist of a reducible* $\underline{3}\oplus\underline{1}$ *representation of* $SU(2)\otimes U(1)$.]

In the conventional Standard Model the breakup of species into generations is inserted "by hand." It is not a consequence of the existence of SU(2)⊗U(1) symmetry or symmetry

[104] They are not the Higgs particles of the SU(2)⊗U(1) ElectroWeak sector.

breaking. In our approach the U(4) Generation group causes the appearance of generations. We base the existence of the Generation group[105] on the four, conserved particle numbers. Leibniz' Principle and Ockham's Razor then lead to the above construction/derivation.

16.5 Generation Group Higgs Mechanism for Fermion Masses

We now consider the Generation group Higgs Mechanism for the eight species of fermions (four species of "normal" matter[106] and four species of Dark Matter). We shall consider the mass terms for the four normal species, which is the same as that of the four Dark species except for the values in the various species mass matrices. Therefore we define the initial 4-vector for the generations of the normal species by

$$\Psi_s = \begin{bmatrix} \psi_{11} \\ \psi_{12} \\ \psi_{13} \\ \psi_{14} \\ \cdots \\ \psi_{41} \\ \psi_{42} \\ \psi_{43} \\ \psi_{44} \end{bmatrix} \qquad (16.46)$$

where ψ_{ki} is the generation index for the ith generation of the kth species. ψ_{k1} is the wave function for the 1st generation, ψ_{k4} is the 4th generation member of the kth species, and we omit other indices in the interests of clarity. The normal fermion species order here are: charged lepton (k = 1), up-type quark, neutral lepton, and down-type quark (k = 4). Other indices of these wave functions are suppressed in the interests of clarity. A 4th generation fermion of any species is yet to be found experimentally. The lagrangian density mass terms for the four normal fermion species are

$$\mathcal{L}_U^{\text{UHiggs}}{}_{\text{FermionMasses}} = \Sigma_{k,\alpha,\beta}\, \bar{\psi}_{kL\alpha}\, \eta_k m_{k\alpha\beta} \psi_{kR\beta} + \text{c.c.} \qquad (16.47)$$

where $m_{k\alpha\beta}$ is complex constant matrix, where k labels species, and where $\alpha, \beta = 1, \ldots, 4$. The total of fermion lagrangian mass terms is

---

[105] In earlier books we suggested the fermion generations might be the result of a wormhole to another 4-dimensional universe. The new approach is simpler and more consistent with known facts – thus more consistent with the Leibniz Minimax Principle.

[106] Not taking account of the three color quark species of normal matter yet.

$$\mathcal{L}^{Higgs}{}_{FermionMasses} = \mathcal{L}_U{}^{UHiggs}{}_{FermionMasses} + \mathcal{L}_{EW}{}^{Higgs}{}_{FermionMasses} \qquad (16.48)$$

where $\mathcal{L}_{EW}{}^{Higgs}$ is the contribution of ElectroWeak Higgs Mechanism to the fermion masses (discussed in the following chapter). Using the vacuum expectation value of η in eq. 16.37 we find

$$\mathcal{L}_U{}^{UHiggs}{}_{FermionMasses} = \Sigma_{\alpha,\beta} \{\bar{\psi}_{2L\alpha}\, \rho_1 m_{2\alpha\beta} \psi_{2R\beta} + \bar{\psi}_{4L\alpha}\, \rho_2 m_{4\alpha\beta} \psi_{4R\beta}\} + \text{c.c.} \qquad (16.49)$$

giving mass terms for the up-type and down-type quark species but not for lepton species. There is an implicit color summation over the color quarks in each generation and quark species. *Qualitatively eq. 16.49 could be viewed as corresponding to the experimentally known largeness of quark masses relative to lepton masses in each generation of normal matter.*

The mass matrices $m_2 = [m_{2\alpha\beta}]$ and $m_4 = [m_{4\alpha\beta}]$ are both complex, constant mass matrices. They can be brought to diagonal form with non-negative values by U(4) matrices A_k and B_k:

$$\begin{aligned} A_2 m_2 B_2{}^{-1} &= D_2 \\ A_4 m_4 B_4{}^{-1} &= D_4 \end{aligned} \qquad (16.50)$$

or

$$\begin{aligned} m_2 &= A_2{}^{-1} D_2\, B_2 \\ m_4 &= A_4{}^{-1} D_4\, B_4 \end{aligned} \qquad (16.51)$$

We now note, that although, both D_2 and D_4 have non-negative real values, down-type quarks are all tachyonic and up-type quarks are all non-tachyonic due to their lagrangian kinetic terms as seen in chapter 5.

We further note that $m_2{}^\dagger m_2$ and $m_4{}^\dagger m_4$ are hermitean, and A_k and B_k are members of U(4) as is D_k for k = 2,4, with the result that m_2 and m_4 are also both members of the U(4) group. Thus

$$\begin{aligned} m_2{}^{-1} &= m_2{}^\dagger \\ m_4{}^{-1} &= m_4{}^\dagger \end{aligned} \qquad (16.52)$$

We can express the mass matrices in terms of U(4) generators

$$\begin{aligned} m_2 &= \Sigma G_i m_{2i} \\ m_4 &= \Sigma G_i m_{4i} \end{aligned} \qquad (16.53)$$

$$\begin{aligned} m_2{}^{-1} &= m_2{}^\dagger = \Sigma G_i m_{2i}{}^* \\ m_4{}^{-1} &= m_4{}^\dagger = \Sigma G_i m_{4i}{}^* \end{aligned} \qquad (16.54)$$

since the matrices G_i are all hermitean, where $\{m_{2i}\}$ and $\{m_{4i}\}$ are each a set of sixteen complex constants.

While we do not as yet know the 4th generation fermions or their masses, the third generation quarks have masses that are far greater than the 1st and 2nd generation quarks or their sum suggesting that the trace of m_2 and m_4.is dominated by the 4th generation mass of the two quark species with a similar situation holding, perhaps, for the two Dark quark species. Therefore if we take the trace of m_2 and m_4 then it seems probable based on the trend of the generations that the 4th generation mass dominates the trace:

$$D_{24} \approx \text{tr } D_2 \qquad (16.55)$$
$$D_{44} \approx \text{tr } D_4$$

We can use these A_k and B_k U(4) transformations to define the eight "physical" (up to further ElectroWeak Higgs Mehanism effects) up-type and down-type quark generations fields:

$$\bar{\psi}_{2L\alpha}\, \rho_1 m_{2\alpha\beta} \psi_{2R\beta} + \bar{\psi}_{4L\alpha}\, \rho_2 m_{4\alpha\beta} \psi_{4R\beta} = (\bar{\psi}_{2L} A_2^{-1})_\alpha \rho_1 D_{2\alpha\beta} (B_2 \psi_{2R})_\beta + (\bar{\psi}_{4L} A_4^{-1})_\alpha \rho_2 D_{4\alpha\beta} (B_4 \psi_{4R})_\beta$$
$$= \bar{\psi}_{2Lphys\alpha}\, \rho_1 D_{2\alpha\beta} \psi_{2Rphys\beta} + \bar{\psi}_{4Lphs\alpha}\, \rho_2 D_{4\alpha\beta} \psi_{4Rphys\beta} \qquad (16.56)$$

Species: up-type quarks down-type quarks

The preceding discussion with changes in the values of constants and constant matrices holds for Dark Matter also where the Dark quarks acquire mass terms but the Dark leptons do not. The Dark Matter species mass terms, with the subscript D signifying Dark Matter, are

$$= \bar{\psi}_{D2Lphys\alpha}\, \rho_{D1} D_{D2\alpha\beta} \psi_{D2Rphys\beta} + \bar{\psi}_{D4Lphs\alpha}\, \rho_{D2} D_{D4\alpha\beta} \psi_{D4Rphys\beta} \qquad (16.57)$$

Dark Species: up-type quarks down-type quarks

END OF EXTRACT FROM Blaha (2015a) and (2015c)

11.3.1 Form of the Generation Group Interaction

The U(4) Generation group generators are denoted G_i and its gauge fields are denoted $U_{\mu i}(X)$. Thus the Generation group term in covariant derivatives is

$$g_G \mathbf{U}_\mu \cdot \mathbf{G} \qquad (11.51)$$

by the discussions in section 16.1 above.

11.4 Layer Group Interactions

The Layer group has been presented in prior books. Below are extracts from those books that describe the origin and features of the Layer group. Section 11.4.1 specifies the Layer group interaction term in the covariant derivative used to determine the Riemann-Christoffel curvature tensor. Fig. 11.1 below displays the fermion layers due to the Layer group. The discussion in the next abstract is based on the fermion layers 'periodic table' of Fig. 11.1.

EXTRACT FROM Blaha (2016b)

2. Equipartition Principle for Fermion Degrees of Freedom

In a closed system at equilibrium the thermal energy of a system is equally partitioned (distributed) among its degrees of freedom. This Equipartition Principle is well known. The application of this principle to the beginning of the universe when all fermions were massless and all symmetries were unbroken suggests that the distribution of fermions should be the same for all fermionic degrees of freedom at that time. Thus there should be approximately equal numbers of fermions of each of the 192 types (See Fig. 11.1 below.) with the same fraction of the total thermal energy.

In the next chapter we will estimate the relative proportion of Normal and Dark matter in the universe at its beginning based on the Equipartition Principle.

6. Proportion of Dark Matter in the Universe

In the beginning, if the Equipartition Principle applies to the 192 massless fermions we can estimate the proportion of the matter in the universe that is Dark. (See Fig. 11.1 below.)

First we note that 8 of the 12 species in layer 1 – the layer with which we are familiar are Normal matter fermions. (Our discussion is based on our Extended Standard Model.) Four of the 12 first layer species are Dark.

The other three layers are all Dark from our point of view since they have not been detected. Thus we find that 40 of the 48 species are Dark yielding a percentage of Dark matter equal to 40/48 = 83.33%. The same count could have been performed by counting fermions with the same results.

Recent studies of the proportion of Dark matter in the universe have yielded two estimates: 84.5% by Aghanim et al in Astronomy and Astrophysics 1303; 5062 and 81.5% from a NASA fit to various models.

Thus our estimate based on our fermion Equipartition Principle is midway between these experimental estimates.

Two possibilities emerge with respect to the present proportion of Dark Matter:

1. The percentage has not changed from the Beginning and the approximate estimates are slightly off. The lack of change could be due to the extremely small decay rates of the

fermions in the higher levels.

2. The percentage of matter in the upper layers has decreased due to decay (See the previous chapter's discussion of the new interaction.) and so the current proportion may be somewhat below 83.33%.

The following page shows the Dark fermion part of the fermion periodic table.

7. Penetrating the veil of the Big Bang

Studies of the Big Bang are hampered by infinities that often appear at t = 0. There appear to be two possible initial states for the expansion of our universe. One possibility is that the universe existed in a metastable? state for some time before expansion began. The other possibility is instantaneous creation. We will not consider this possibility although it could support the Equipartition Principle. We have previously suggested a model (that includes inflation) that has an extremely short initial metastable state.[12]

If there was an initial period of quasi-stability, then it is possible that a state of equilibrium was achieved and the fermion Equipartition Principle applied. Then the estimate of the proportion of Dark matter would be correct.

Today experiments at CERN and other accelerators are creating quark-gluon plasma states that approximate the conditions near the Big Bang point. It would be interesting if it were possible to determine the proportions of all fermions generated in an ion-ion collision – both leptons and quarks – to see if the numbers of each type of produced fermion are approximately equal. If a rough equality were found then there would be significant support for the fermion Equipartition Principle.

If the Equipartition Principle holds at the time of the Big Bang then modeling of the Big Bang state would be constrained to be more realistic and we will have begun to penetrate the veil of the Big Bang.

Layer 4

Layer 3

Layer 2

Layer 1 (Our Layer)

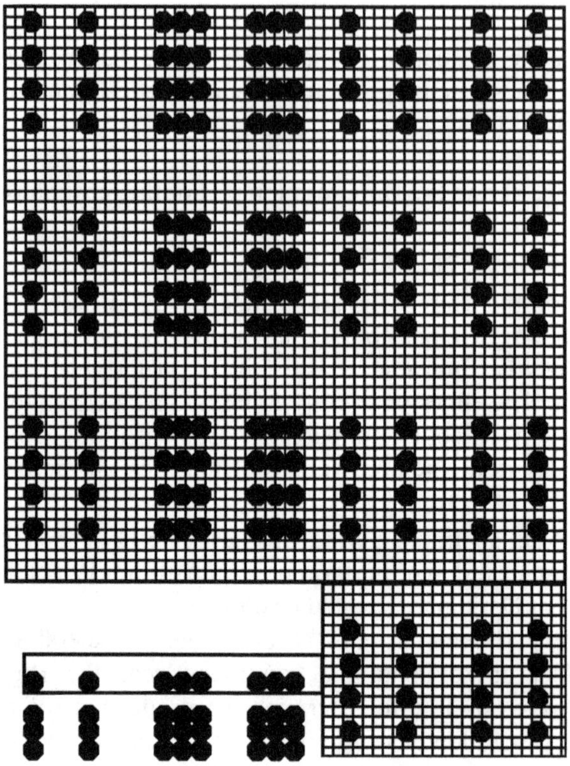

Figure 11.1. Dark parts of the periodic table are 'cross-hatched.' Light parts are the known fermions with an additional, as yet not found, 4th generation of layer 1 is shown boxed. It is part of Dark matter at present. When found experimentally it will be 'non-Dark.'

END OF EXTRACT FROM Blaha (2016b)

EXTRACT FROM Blaha (2016c)

Chapter 4. The Broken U(4) Layer Group and Fermion Layers

4.1 The Layer Group

The Generation group was based on U(4) rotations of the four number operators B, L, B_D, and L_D in the one generation Extended Standard Model. We can visualize these rotations horizontally as

Given U(4) in the one generation case, it is natural to assume that the symmetry applies in a larger case – the case of a four generation Extended Standard Model. We attach a U(4) vector generation index to each fermion field ranging from 1 through 4 and introduce interactions between the generations. The result is the four generation Extended Standard Model presented in Blaha (2015a) and earlier books.

After generations are introduced, then it becomes possible to consider vertical rotations amongst the four generations:

L_1
L_2
L_3
L_4

These rotations are based on four conserved numbers L_1, L_2, L_3, and L_4 that count the number of fermions of each generation in a state.[107] L_i counts the number of fundamental fermions in generation i for i = 1, 2, 3, 4. Fermions have positive $L_i = +1$ values and anti-fermions have negative $L_i = -1$ values. For example, if a state has 3 u quarks, 1 d quark, 1 anti-s quark, 2 electrons, 2 anti-τ leptons, 2 Dark muon neutrinos, and one Dark electron neutrino ν_{De} then $L_1 = 3+1+2+1 = 7$, $L_2 = -1+2 = 1$, $L_3 = -2$, and $L_4 = 0$.

It is important to note that the Layer numbers are independent of the baryon and lepton particle numbers that form the basis of the Generation group, and so the physics embodied in the Generation group is not the same as the physics of the Layer group defined below.

Layer numbers are conserved under strong and electromagnetic interactions but broken by the ElectroWeak interactions.

The partial conservation of the L_i Numbers enables us to define a new broken U(4) symmetry called the Layer group. Thus we have four new (partly) conserved particle numbers. Linear combinations of these numbers are also (partly) conserved:[108]

$$L_1' = aL_1 + bL_2 + cL_3 + dL_4$$
$$L_2' = eL_1 + fL_2 + gL_3 + hL_4$$
$$L_3' = iL_1 + jL_2 + kL_3 + lL_4$$
$$L_4' = mL_1 + nL_2 + oL_3 + pL_4$$

using constants labeled from "a" through "p" or

$$L' = AL$$

[107] The generations are numbered from 1 to 4 with the lowest masses generation (e, ν_e, u, d) being generation 1.
[108] This discussion parallels the discussion of particle numbers in the following Extract.

where A is a 4×4 U(4) matrix.

If we want the new 'primed' set of conserved numbers to be an independent set of numbers then the determinant of the constants must be non-zero. Thus the matrix, A, is invertible.

The set of 4×4 matrices of the above type form a U(4) group.[109] *The choice of U(4) rather than SU(4) is required since there are four independent layer particle numbers and U(4) has four diagonal matrices in its algebra while SU(4) only has three diagonal matrices. U(4) preserves the independence of the four independent particle numbers. Thus we have the Layer group.*

Since the coefficients in Layer group transformations can be local functions, the Layer group is implemented as a Yang-Mills theory.

Just as we extended the reach of the Generation group from one generation to four generations, we can extend the Layer group similarly by assuming that there are four layers of fermions by adding a U(4) vector layer index to each fermion field. Further, since we only know of one layer – our layer – the gauge fields for all vector interactions must be different[110] for each layer. The only exception being gravitation which is universal to all layers. Other layers must then be "Dark" in the sense that each layer is independent except for a new ultra-weak interaction that connects states of different layers.

To implement the Layer symmetry in the Extended Standard Model lagrangian the following steps are required:

1, All covariant derivatives must acquire another interaction term with 16 U(4) fields corresponding to the 16 generators of the U(4) Layer group.

2. Expand the Theory of Everything to embody the Layer group symmetry by adding another index ranging from 1 through 4 to each fermion field making four layers of four generations of fermions. Thus the fermion fields are in the fundamental representation of the Layer group.

3. Expand the Theory of Everything by adding a Layer group index to each vector gauge field so that each layer has its own complete set of SU(3)⊗SU(2)⊗U(1)⊗U(4)⊗U(4)⊗U(4) gauge fields.[111] The General Relativity Reality group

---

[109] If we wish to further limit the values of the 'primed' Numbers to integers assuming the unprimed Numbers are integers then the group of the transformation is the set of permutations of four entities – the Symmetric group S_4. However the 'primed' numbers may be integers or not. There is no apparent physical principle requiring integer Numbers. Quarks are usually assigned Baryon Number 1/3. Also Numbers are not necessarily positive valued.

[110] Later we will see that mass mixing occurs between particles in different layers, which in turn leads to mixing of the gauge fields of the various layers. However the ultra-weakness of the Layer gauge field interactions makes the gauge field mixing negligible – although it is present in principle.

[111] This symmetry group product was changed in this book from the original in this excerpt.

U(4) gauge interactions are commonly experienced by all layers. Add Layer indexes to all gauge field dynamic lagrangian terms.

4. Each layer should have its own set of gauge field Higgs particles (modulo miniscule mixing) for other gauge field interactions. One expects that the masses of fermions and gauge fields should be substantially larger for the three 'upper' layers beyond our layer. Otherwise we would have found particles from these upper layers.

5. Insert an interaction term of the form $g_v G_L{}^a V^a{}_\mu$ in the covariant derivative of each fermion using the 16 U(4) Layer gauge fields $V^a{}_\mu$ with a = 1, 2, ..., 16 and the 16 U(4) generators $G_L{}^a$ that couple within and between layers using the new layer index where g_v is an ultra-weak coupling constant much smaller than the ElectroWeak coupling constants. This interaction will be between normal matter, between Dark Matter, between normal and Dark matter, and between fermions in the same and different layers. This interaction will be the only interaction between particles in different layers. Its small coupling constant and presumably very large gauge field masses will essentially make the four layers almost independent of each other except for gravitation.

6. Insert Layer group symmetry breaking Higgs fields (independently for all layers) that generate fermion and gauge field mass term contributions independently for each layer.

The above modifications to the Theory of Everything lagrangian will be mathematically similar to the development of the features of the Generation group in Blaha (2015a) presented in the extract below.

The form of the "periodic table" of fermions that results appears in Fig. 11.1 above.

4.2 Layer Group Interactions

The new Layer group interactions play two roles:

1. The four diagonal generator terms of $U_L{}^a V^a{}_\mu$ create transitions amongst the Normal and Dark fermions, including transitions between Normal and Dark, between Normal and Normal, and between Dark and Dark in each layer separately.

2. The non-diagonal generator terms of $U_L{}^a V^a{}_\mu$ create transitions amongst the Normal and Dark fermions between *different* layers. The non-diagonal interactions are the source of decays from higher layers to lower layers. Since no experimental evidence of such decays has been found as yet we conclude that the interaction constant g_V is ultra-small and/or the Layer gauge boson masses are extraordinarily large. Thus the distribution of fermions in layers may be approximately the same as that at the time of the Big Bang. We considered this issue in Blaha (2016a) and (2016b), and showed the Dark Matter fraction implied by the fermion spectrum (Fig. 3.2) is 83.33% – a result consistent with cosmological data. The

fermions of the higher layers must be constituents of "Dark Matter" (as well as the Dark part of our layer) since only the ultra-weak Layer interaction and graviton interactions can connect layers.

...

4.4 Layer Group Interaction

In this section we describe Layer group interactions.[112] The Theory of Everything symmetry group that we have developed in earlier books was:

$$SU(3)\otimes SU(2)\otimes U(1)\otimes SU(2)\otimes U(1)\otimes U(4)\otimes U(4)) \tag{4.1}$$

Each factor in this product is a subgroup of U(8) but the product of the subgroups is *not* a subgroup of U(8) because the elements of each factor does not commute with the elements of other factor groups. The total number of independent generators of the product

$$SU(3)\otimes SU(2)\otimes U(1)\otimes SU(2)\otimes U(1)\otimes U(4)\otimes U(4)$$

is 48. The total number of generators of U(8) is 64. The difference, 16 generators, constitutes the generators of another U(4) group. We therefore propose that this U(4) group, which we call the Layer group, be a part of the Theory of Everything group yielding:

$$SU(3)\otimes SU(2)\otimes U(1)\otimes U(4)\otimes U(4)\otimes U(4) \tag{4.2}$$

Its role will be to provide an ultra-weak interaction between the known normal fermions and the Dark fermions. In addition the Layer group also naturally leads to four layers of fermions. Our normal matter and its associated Dark Matter we call Layer 1. The other three layers, with presumably very high masses, are Dark Matter also and remain to be discovered.

We now modify the leptonic left-handed and right-handed covariant derivatives in the normal and Dark ElectroWeak sectors, which were:[113]

$$D_{NL\mu} = \partial/\partial X^\mu - \tfrac{1}{2}ig\boldsymbol{\sigma}\cdot\mathbf{W}_\mu + \tfrac{1}{2}ig'B_\mu - \tfrac{1}{2}ig_G\mathbf{G}\cdot\mathbf{U}_\mu \tag{4.3}$$
$$D_{DL\mu} = \partial/\partial X^\mu - \tfrac{1}{2}ig_D\boldsymbol{\sigma}\cdot\mathbf{W}'_\mu + \tfrac{1}{2}ig_D'B'_\mu + \tfrac{1}{2}ig_D''B_\mu - \tfrac{1}{2}ig_G\mathbf{G}\cdot\mathbf{U}_\mu \tag{4.4}$$
$$D_{NR\mu} = \partial/\partial X^\mu + ig'B_\mu - \tfrac{1}{2}ig_G\mathbf{G}\cdot\mathbf{U}_\mu \tag{4.5}$$
$$D_{DR\mu} = \partial/\partial X^\mu + \tfrac{1}{2}ig_D'B'_\mu + \tfrac{1}{2}ig_D''B_\mu - \tfrac{1}{2}ig_G\mathbf{G}\cdot\mathbf{U}_\mu \tag{4.6}$$

[112] Much of this section is extracted from Blaha (2016b).
[113] Equations from Blaha (2015a).

where the B_μ term in eqs. 4.4 and 4.6 above provide a transition between normal and dark matter. We also now modify eqs. 4,8 and 4.10 in the quark ElectroWeak covariant derivatives, which were:

$$D_{qNL\mu} = \partial/\partial X_c^\mu - \tfrac{1}{2}ig\boldsymbol{\sigma}\cdot\mathbf{W}_\mu - ig'B_\mu/6 + ig_C\tau\cdot A_{C\mu} - \tfrac{1}{2}ig_G\mathbf{G}\cdot\mathbf{U}_\mu \quad (4.7)$$
$$D_{qDL\mu} = \partial/\partial X_c^\mu - \tfrac{1}{2}ig_D\boldsymbol{\sigma}\cdot\mathbf{W}'_\mu + \tfrac{1}{2}ig_D'B'_\mu + \tfrac{1}{2}ig_D''B_\mu - \tfrac{1}{2}ig_G\mathbf{G}\cdot\mathbf{U}_\mu \quad (4.8)$$
$$D_{qNR\mu} = \partial/\partial X_c^\mu + ig'B_\mu/3 + ig_C\tau\cdot A_{C\mu} - \tfrac{1}{2}ig_G\mathbf{G}\cdot\mathbf{U}_\mu \quad (4.9)$$
$$D_{qDR\mu} = \partial/\partial X_c^\mu + \tfrac{1}{2}ig_D'B'_\mu + \tfrac{1}{2}ig_D''B_\mu - \tfrac{1}{2}ig_G\mathbf{G}\cdot\mathbf{U}_\mu \quad (4.10)$$

where $A_{C\mu}$ is the color gauge field.

The new covariant derivatives that contain the U(4) Layer group interaction are:

1) Leptonic covariant derivatives in the normal and Dark ElectroWeak sectors:

$$D_{NL\mu} = \partial/\partial X^\mu - \tfrac{1}{2}ig\boldsymbol{\sigma}\cdot\mathbf{W}_\mu + \tfrac{1}{2}ig'B_\mu - \tfrac{1}{2}i\, g_v\mathbf{G}_L\cdot\mathbf{V}_\mu - \tfrac{1}{2}ig_G\mathbf{G}\cdot\mathbf{U}_\mu \quad (4.3')$$
$$D_{DL\mu} = \partial/\partial X^\mu - \tfrac{1}{2}ig_D\boldsymbol{\sigma}\cdot\mathbf{W}'_\mu + \tfrac{1}{2}ig_D'B'_\mu - \tfrac{1}{2}i\, g_v\mathbf{G}_L\cdot\mathbf{V}_\mu - \tfrac{1}{2}ig_G\mathbf{G}\cdot\mathbf{U}_\mu \quad (4.4')$$
$$D_{NR\mu} = \partial/\partial X^\mu + ig'B_\mu - \tfrac{1}{2}i\, g_v\mathbf{G}_L\cdot\mathbf{V}_\mu - \tfrac{1}{2}ig_G\mathbf{G}\cdot\mathbf{U}_\mu \quad (4.5')$$
$$D_{DR\mu} = \partial/\partial X^\mu + \tfrac{1}{2}ig_D'B'_\mu - \tfrac{1}{2}i\, g_v\mathbf{G}_L\cdot\mathbf{V}_\mu - \tfrac{1}{2}ig_G\mathbf{G}\cdot\mathbf{U}_\mu \quad (4.6')$$

2) Quark ElectroWeak covariant derivatives in the normal and Dark ElectroWeak sectors:

$$D_{qNL\mu} = \partial/\partial X_c^\mu - \tfrac{1}{2}ig\boldsymbol{\sigma}\cdot\mathbf{W}_\mu - ig'B_\mu/6 + ig_C\tau\cdot A_{C\mu} - \tfrac{1}{2}i\, g_v\mathbf{G}_L\cdot\mathbf{V}_\mu - \tfrac{1}{2}ig_G\mathbf{G}\cdot\mathbf{U}_\mu \quad (4.7')$$
$$D_{qDL\mu} = \partial/\partial X_c^\mu - \tfrac{1}{2}ig_D\boldsymbol{\sigma}\cdot\mathbf{W}'_\mu + \tfrac{1}{2}ig_D'B'_\mu - \tfrac{1}{2}i\, g_v\mathbf{G}_L\cdot\mathbf{V}_\mu - \tfrac{1}{2}ig_G\mathbf{G}\cdot\mathbf{U}_\mu \quad (4.8')$$
$$D_{qNR\mu} = \partial/\partial X_c^\mu + ig'B_\mu/3 + ig_C\tau\cdot A_{C\mu} - \tfrac{1}{2}i\, g_v\mathbf{G}_L\cdot\mathbf{V}_\mu - \tfrac{1}{2}ig_G\mathbf{G}\cdot\mathbf{U}_\mu \quad (4.9')$$
$$D_{qDR\mu} = \partial/\partial X_c^\mu + \tfrac{1}{2}ig_D'B'_\mu - \tfrac{1}{2}i\, g_v\mathbf{G}_L\cdot\mathbf{V}_\mu - \tfrac{1}{2}ig_G\mathbf{G}\cdot\mathbf{U}_\mu \quad (4.10')$$

We add a new Layer group index to each fermion field[114] with index number values ranging from 1 through 4. The new Layer group interaction term $g_vG_L{}^aV^a{}_\mu$ uses 16 U(4) gauge fields $V^a{}_\mu$ with a = 1, 2, …, 16, and 16 U(4) generators, denoted $G_L{}^a$, *that couple to the new Layer group indexes.*[115] The $V^a{}_\mu$ gauge fields have a standard U(4) kinetic energy lagrangian term.

The Layer interaction constant g_v is an ultra-weak coupling constant assumed to be much smaller than the ElectroWeak and Generation group coupling constants.

In addition all gauge vector bosons acquire a Layer group index appropriate to the layer and all Higgs bosons of each layer have a layer index appropriate to the layer. Thus each layer

[114] Gauge bosons and Higgs bosons also acquire Layer group indexes. Thus each layer is entirely self-contained with the only interactions between layers being Layer group interactions and gravitation.

[115] The Generation group U(4) matrices G^a couple to Generation group indices.

is effectively self-contained,[116] with different fermion and boson particle masses, with only the ultra-weak Layer interaction and gravitation coupling layers.

4.5 Layer Group Higgs Mechanism Contributions to Layer Gauge Field Masses

In this section we will determine the Layer group Higgs contributions to gauge field masses. (The fermion mass contributions from the various Higgs interactions are shown in eq. 5.56' in chapter 3. We will see that all[117] layers have Layer group Higgs contributions to their fermion masses.)

We begin by assuming that a scalar Higgs field η exists, which is a U(4) Layer group 4-vector with four components corresponding to the four conserved generation number operators L_1, L_2, L_3, and L_4 of section 4.1. η is an SU(2)⊗U(1)⊗SU(3) ElectroWeak and Strong Interaction scalar. Its lagrangian density terms are[118]

$$\mathcal{L}_V^{\text{Higgs}} = (\partial\eta^\dagger/\partial X^\mu)(\partial\eta/\partial X^\mu) - \lambda(\eta^\dagger\eta - \rho^2)^2 + \mathcal{L}_V^{\text{Higgs}}{}_{\text{FermionMasses}} \tag{4.11}$$

where $\mathcal{L}_V^{\text{Higgs}}{}_{\text{FermionMasses}}$ are the fermion masses produced by the Layer Higgs Mechanism and where we set the η Layer 4-vector with Higgs field components to

$$\eta = \begin{bmatrix} \rho_1 \\ \rho_2 \\ \rho_3 \\ \rho_4 \end{bmatrix} \qquad \begin{array}{c} \underline{\text{Corresponding Conserved Number}} \\ L_1 \\ L_2 \\ L_3 \\ L_4 \end{array} \tag{4.12}$$

where ρ_1, ρ_2, ρ_3 and ρ_4 are real fields.[119] Then the covariant derivative of η is

$$D_{\dots\mu}\eta = \{\partial/\partial X^\mu + \dots - \tfrac{1}{2}ig_V[\Sigma \mathbf{G}_{Li}\mathbf{V}_{i\mu} + \mathbf{G}_{L1}\mathbf{V}_{1\mu} + \mathbf{G}_{L2}\mathbf{V}_{2\mu} + \mathbf{G}_{L3}\mathbf{V}_{3\mu} + \mathbf{G}_{L4}\mathbf{V}_{4\mu}]\} \begin{bmatrix} \rho_1 \\ \rho_2 \\ \rho_3 \\ \rho_4 \end{bmatrix} \tag{4.13}$$

---

[116] Modulo mass matrix mixing between the layers that will modify the fermion spectrum and gauge fields. These modifications are assumed to be very small due to the ultra-weak nature of the Layer group interaction.

[117] All Layers have Layer group Higgs contributions is required to avoid massless Layer group gauge fields.

[118] Again we use the standard formulation of the Higgs Mechanism because of its familiarity.

[119] Each field ρ_i can be expressed as a pseudoquantum field: $\rho_i = \varphi_{1i} + \varphi_{2i}$ where φ_{1i} has the vacuum expectation value $\rho_{i0.}$ for i = 1, … , 4. Thus our pseudoquantum field theory version is implemented easily.

The sum over i is from 5 through 16 (non-diagonal matrices), and $[\mathbf{G}_{Li}]_{jk}$ is the jkth element of $\mathbf{G}_{Li}$. Then

$$D_{\ldots\mu}\eta = \begin{bmatrix} \partial\rho_1/\partial X^\mu - \tfrac{1}{2} i g_V\{\rho_1 \mathbf{G}_{L1}\mathbf{V}_{1\mu} + \rho_2\Sigma[\mathbf{G}_{Li}]_{11}\mathbf{V}_{i\mu} + \rho_2\Sigma[\mathbf{G}_{Li}]_{12}\mathbf{V}_{i\mu} + \rho_3\Sigma[\mathbf{G}_{Li}]_{13}\mathbf{V}_{i\mu} + \rho_4\Sigma[\mathbf{G}_{Li}]_{14}\mathbf{V}_{i\mu}\} \\ \partial\rho_2/\partial X^\mu - \tfrac{1}{2} i g_V\{\rho_2 \mathbf{G}_{L2}\mathbf{V}_{2\mu} + \rho_1\Sigma[\mathbf{G}_{Li}]_{21}\mathbf{V}_{i\mu} + \rho_2\Sigma[\mathbf{G}_{Li}]_{22}\mathbf{V}_{i\mu} + \rho_3\Sigma[\mathbf{G}_{Li}]_{23}\mathbf{V}_{i\mu} + \rho_4\Sigma[\mathbf{G}_{Li}]_{24}\mathbf{V}_{i\mu}\} \\ \partial\rho_3/\partial X^\mu - \tfrac{1}{2} i g_V\{\rho_3 \mathbf{G}_{L3}\mathbf{V}_{3\mu} + \rho_1\Sigma[\mathbf{G}_{Li}]_{31}\mathbf{V}_{i\mu} + \rho_2\Sigma[\mathbf{G}_{Li}]_{32}\mathbf{V}_{i\mu} + \rho_3\Sigma[\mathbf{G}_{Li}]_{33}\mathbf{V}_{i\mu} + \rho_4\Sigma[\mathbf{G}_{Li}]_{34}\mathbf{V}_{i\mu}\} \\ \partial\rho_4/\partial X^\mu - \tfrac{1}{2} i g_V\{\rho_4 \mathbf{G}_{L4}\mathbf{V}_{4\mu} + \rho_1\Sigma[\mathbf{G}_{Li}]_{41}\mathbf{V}_{i\mu} + \rho_2\Sigma[\mathbf{G}_{Li}]_{42}\mathbf{V}_{i\mu} + \rho_3\Sigma[\mathbf{G}_{Li}]_{43}\mathbf{V}_{i\mu} + \rho_4\Sigma[\mathbf{G}_{Li}]_{44}\mathbf{V}_{i\mu}\} \end{bmatrix} \tag{4.14}$$

$$= \begin{bmatrix} \partial\rho_1/\partial X^\mu - \tfrac{1}{2} i g_V\{\rho_1 \mathbf{G}_{L1}\mathbf{V}_{1\mu} + \rho_2\Sigma[\mathbf{G}_{Li}]_{12}\mathbf{V}_{i\mu} + \rho_3\Sigma[\mathbf{G}_{Li}]_{13}\mathbf{V}_{i\mu} + \rho_4\Sigma[\mathbf{G}_{Li}]_{14}\mathbf{V}_{i\mu}\} \\ \partial\rho_2/\partial X^\mu - \tfrac{1}{2} i g_V\{\rho_2 \mathbf{G}_{L2}\mathbf{V}_{2\mu} + \rho_1\Sigma[\mathbf{G}_{Li}]_{21}\mathbf{V}_{i\mu} + \rho_3\Sigma[\mathbf{G}_{Li}]_{23}\mathbf{V}_{i\mu} + \rho_4\Sigma[\mathbf{G}_{Li}]_{24}\mathbf{V}_{i\mu}\} \\ \partial\rho_3/\partial X^\mu - \tfrac{1}{2} i g_V\{\rho_3 \mathbf{G}_{L3}\mathbf{V}_{3\mu} + \rho_1\Sigma[\mathbf{G}_{Li}]_{31}\mathbf{V}_{i\mu} + \rho_2\Sigma[\mathbf{G}_{Li}]_{32}\mathbf{V}_{i\mu} + \rho_4\Sigma[\mathbf{G}_{Li}]_{34}\mathbf{V}_{i\mu}\} \\ \partial\rho_4/\partial X^\mu - \tfrac{1}{2} i g_V\{\rho_4 \mathbf{G}_{L4}\mathbf{V}_{4\mu} + \rho_1\Sigma[\mathbf{G}_{Li}]_{41}\mathbf{V}_{i\mu} + \rho_2\Sigma[\mathbf{G}_{Li}]_{42}\mathbf{V}_{i\mu} + \rho_3\Sigma[\mathbf{G}_{Li}]_{43}\mathbf{V}_{i\mu}\} \end{bmatrix} \tag{4.15}$$

since the generators $\mathbf{G}_i$ have zeroes along their diagonals for $i = 5, \ldots, 16$.

From eq. 4.15 we find the corresponding Higgs field kinetic terms in the lagrangian are

$$(D_{\ldots\mu}\eta)^\dagger D^{\mu}_{\ldots}\eta = \partial\rho_1/\partial X^\mu\, \partial\rho_1/\partial X_\mu + \partial\rho_2/\partial X^\mu\, \partial\rho_2/\partial X_\mu + \partial\rho_3/\partial X^\mu\, \partial\rho_3/\partial X_\mu + \partial\rho_4/\partial X^\mu\, \partial\rho_4/\partial X_\mu + \\ + g_V^2\rho_1^2\mathbf{V}_{1\mu}\mathbf{V}_1^{\mu}/4 + g_V^2\rho_2^2\mathbf{V}_{2\mu}\mathbf{V}_2^{\mu}/4 + g_V^2\rho_3^2\,\mathbf{V}_{3\mu}\mathbf{V}_3^{\mu}/4 + g_V^2\rho_4^2\,\mathbf{V}_{4\mu}\mathbf{V}_4^{\mu}/4 + \ldots \tag{4.16}$$

We now turn to calculating the remaining terms in eq. 4.16 that determine the masses of the remaining 14 gauge fields. We begin by assigning matrix elements for the remaining hermitean U(4) generators:

$$\begin{aligned} [G_{L5}]_{ik} &= \delta_{i1}\delta_{k2} + \delta_{i2}\delta_{k1} \\ [G_{L6}]_{ik} &= -i\delta_{i1}\delta_{k2} + i\delta_{i2}\delta_{k1} \\ [G_{L7}]_{ik} &= \delta_{i1}\delta_{k3} + \delta_{i3}\delta_{k1} \\ [G_{L8}]_{ik} &= -i\delta_{i1}\delta_{k3} + i\delta_{i3}\delta_{k1} \\ [G_{L9}]_{ik} &= \delta_{i1}\delta_{k4} + \delta_{i4}\delta_{k1} \\ [G_{L10}]_{ik} &= -i\delta_{i1}\delta_{k4} + i\delta_{i4}\delta_{k1} \\ [G_{L11}]_{ik} &= \delta_{i2}\delta_{k3} + \delta_{i3}\delta_{k2} \\ [G_{L12}]_{ik} &= -i\delta_{i2}\delta_{k3} + i\delta_{i3}\delta_{k2} \\ [G_{L13}]_{ik} &= \delta_{i2}\delta_{k4} + \delta_{i4}\delta_{k2} \\ [G_{L14}]_{ik} &= -i\delta_{i2}\delta_{k4} + i\delta_{i4}\delta_{k2} \\ [G_{L15}]_{ik} &= \delta_{i3}\delta_{k4} + \delta_{i4}\delta_{k3} \\ [G_{L16}]_{ik} &= -i\delta_{i3}\delta_{k4} + i\delta_{i4}\delta_{k3} \end{aligned} \tag{4.17}$$

Then completing eq. 4.16 using eq. 4.15 we find

$$(D_{\ldots\mu}\eta)^\dagger D_{\ldots}^{\ \mu}\eta = \partial\rho_1/\partial X^\mu\, \partial\rho_1/\partial X_\mu + \partial\rho_2/\partial X^\mu\, \partial\rho_2/\partial X_\mu + \partial\rho_3/\partial X^\mu\, \partial\rho_3/\partial X_\mu + \partial\rho_4/\partial X^\mu\, \partial\rho_4/\partial X_\mu +$$
$$+ g_V^2\rho_1^2 \mathbf{V}_{1\mu}\mathbf{V}_1^{\ \mu}/4 + g_V^2\rho_2^2\, V_2^{\ 2}/4 + g_V^2\rho_3^2\, V_3^{\ 2}/4 + g_V^2\rho_4^2\, V_4^{\ 2}/4 +$$
$$+ (g_V/2)^2(\rho_1^{\ 2} + \rho_2^{\ 2})(V_5^{\ 2} + V_6^{\ 2}) + (g_V/2)^2(\rho_1^{\ 2} + \rho_3^{\ 2})(V_7^{\ 2} + V_8^{\ 2}) +$$
$$+ (g_V/2)^2(\rho_1^{\ 2} + \rho_4^{\ 2})(V_9^{\ 2} + V_{10}^{\ 2}) + (g_V/2)^2(\rho_2^{\ 2} + \rho_3^{\ 2})(V_{11}^{\ 2} + V_{12}^{\ 2}) +$$
$$+ (g_V/2)^2(\rho_2^{\ 2} + \rho_4^{\ 2})(V_{13}^{\ 2} + V_{14}^{\ 2}) + (g_V/2)^2(\rho_3^{\ 2} + \rho_4^{\ 2})(V_{15}^{\ 2} + V_{16}^{\ 2})$$

4.18)

up to total divergences, which generate surface terms which we discard, and also assuming that all fields satisfy the gauge condition

$$\partial V_i^{\ \mu}/\partial X^\mu = 0 \tag{4.19}$$

Eq. 4-18 shows all Layer group gauge fields have masses. The combination of an ultra-weak coupling constant and very large gauge field masses results in extremely weak interactions between the fields in each layer, which leads to almost independent layers of normal and Dark fermions. Thus the Darkness! They result in very rare decays between layers, and very weak interactions between fermions in different layers. The higher layers with presumably much more massive fermions are thus well "insulated" from our layer. Thus they are Dark to us as well.

We estimate Layer group gauge field masses to be very large – of the order of many TeV or they would have been detected at CERN by now. Their detection must await the construction of much more powerful accelerators. *The "non-diagonal" Layer gauge fields are the means by which we may hope to eventually find fermions of the higher layers.*

4.6 Layer Group Higgs Mechanism Contributions to Fermion Masses

The fermion masses of the charged lepton, and the up-type quark, and down-type quark species' generations all show a rapid increase of mass with the generation. For example the u quark mass is a few MeV while the t quark (third generation) has a mass of about 170 GeV/c. The ratio of these masses is about 170,000. While one can account for this great difference by the judicious choices of Higgs' parameter values, when one considers the Layer group and its associated numerical quantities: ultra-weak coupling, its very large Layer gauge field masses – perhaps of the order of hundreds or thousands of GeV/c, then a large difference in particle masses between layers is understandable and natural.

The form of the layers of fermion mass terms is[120]

[120] Layer group contributions have been added to the original eq. 5.56 in Blaha (2015b) in accord with Blaha (2016a).

$$\mathcal{L}^{Higgs}{}_{FermionMasses} = \Sigma_{k,a,\alpha,\beta}\bar{\psi}_{kaL\alpha\delta}\,\eta_k m_{EW_{ka\alpha\beta}}\psi_{kaR\beta} + \Sigma_{k,a,\alpha,\beta}\bar{\psi}_{D_{kaL\alpha}}\eta_{Dk} m_{DEW_{ka\alpha\beta}}\psi_{D_{kaR\beta}} + \quad \text{ElectroWeak}$$

$$+ \Sigma_{k,a,\alpha,\beta}\bar{\psi}_{UkaL\alpha}\,\eta_{Uka} m_{Uka\alpha\beta}\psi_{UkaR\beta} + \quad \text{Generation}$$

$$+ \Sigma_{k,a,\alpha,\beta}\bar{\psi}_{DUkaL\alpha}\eta_{DUka} m_{DUka\alpha\beta}\psi_{DUkaR\beta} + \quad \text{Group U}$$

$$+ \Sigma_{k,g,\delta,\gamma}\bar{\psi}_{LkgL\delta}\,\eta_{Lg} m_{Lg\delta\gamma}\psi_{LkgR\gamma} + \quad \text{Layer}$$

$$+ \Sigma_{k,g,\delta,\gamma}\bar{\psi}_{DLkgL\delta}\,\eta_{DLg} m_{DLg\delta\gamma}\psi_{DLkgR\gamma} + \quad \text{Group L}$$

$$+ \Sigma_{k,a}\bar{\psi}_{GkaL}\eta_{Ga} m_{Gka}\psi_{GkaR} + \Sigma_{k,a}\bar{\psi}_{DGkaL}\eta_{DGa} m_{DGka}\psi_{DGkaR} + \quad \text{Gravitational}$$

$$+ \text{c.c.} \tag{5.56'}$$

where the subscripts EW, D, U, L and G label ElectroWeak origin, D Dark type, U Generation group origin, L Layer group origin, and G Gravitational origin respectively. The fields labeled η (with subscripts) are Higgs fields that have non-zero vacuum expectation values.[121] The indices k label species – normal and Dark separately, g labels the (four) generations, and a labels the layers. The indices δ and γ label *layer* rows and columns (with implicit sums over generations in the Layer group terms.) The Layer group mass contribution is the same for each fermion in each generation for each species in each layer. The matrices labeled m (with subscripts) are the complex constant mass matrices of species. The indices α, β = 1, … , 4 label *generation* rows and columns.

Eq. 5.56' contains the mass terms for the four layers of fermions in our Theory of Everything. *For each species and generation the Layer group terms mix the Layer mass contributions.*

$$\mathcal{L}_V{}^{Higgs}{}_{FermionMasses} = \Sigma_{k,g,\delta,\gamma}\bar{\psi}_{LkgL\delta}\,\eta_{Lg} m_{Lg\delta\gamma}\psi_{LkgR\gamma} + \Sigma_{k,g,\delta,\gamma}\bar{\psi}_{DLkgL\delta}\,\eta_{DLg} m_{DLg\delta\gamma}\psi_{DLkgR\gamma} + \text{c.c.} \tag{4.20}$$

where $m_{Lg\delta\gamma}$ and $m_{DLg\delta\gamma}$ are complex constant matrices, and where δ, γ = 1, … , 4 label generations. The total of fermion lagrangian mass terms is

$$\mathcal{L}^{Higgs}{}_{FermionMasses} = \mathcal{L}_{EWFermionMasses} + \mathcal{L}_{UFermionMasses} + \mathcal{L}_{VFermionMasses} + \mathcal{L}_{GravFermionMasses} + \text{c.c.} \tag{4.21}$$

where $\mathcal{L}_{EW}{}^{Higgs}$ is the contribution of ElectroWeak Higgs Mechanism to the fermion masses (discussed in the following chapter). Using the vacuum expectation value of η in eq. 4.12, and assuming $\eta_{Lg} = \eta_{DLg}$ we find

[121] The Higgs fields η… in our pseudoquantum formulation are η… = $\varphi_{1\ldots}(x) + \varphi_{2\ldots}(x)$ as described earlier.

$$\mathcal{L}_{\text{VFermionMasses}} = \Sigma_{k,\delta,\gamma}\{\bar{\psi}_{Lk1L\delta}\rho_1 m_{L1\delta\gamma}\psi_{Lk1R\gamma} + \bar{\psi}_{DLk1L\delta}\rho_1 m_{DL1\delta\gamma}\psi_{DLk1R\gamma} +$$
$$+ \bar{\psi}_{Lk2L\delta}\rho_2 m_{L2\delta\gamma}\psi_{Lk2R\gamma} + \bar{\psi}_{DLk2L\delta}\rho_2 m_{DL2\delta\gamma}\psi_{DLk2R\gamma} +$$
$$+ \bar{\psi}_{Lk3L\delta}\rho_3 m_{L3\delta\gamma}\psi_{Lk3R\gamma} + \bar{\psi}_{DLk3L\delta}\rho_3 m_{DL3\delta\gamma}\psi_{DLk3R\gamma} +$$
$$+ \bar{\psi}_{Lk4L\delta}\rho_4 m_{L4\delta\gamma}\psi_{Lk4R\gamma} + \bar{\psi}_{DLk4L\delta}\rho_4 m_{DL4\delta\gamma}\psi_{DLk4R\gamma}\} + \text{c.c.} \qquad (4.22)$$

where the indices k label species – normal and Dark separately, and the indices δ and γ label *layer* rows and columns. The integers 1, … , 4 label generations.

The mass matrices in eq. 4.21 are complex, constant mass matrices that can be totaled and brought to diagonal form with non-negative values by U(4) matrices. The resulting diagonalized mass matrices are the mass matrices of the physical fermions. See section 4.3 for an example of the procedure.

The fermion masses in the resulting three "upper" layers have terms with similar forms but with different mass values. These values are presumably very large. We expect that they are in the multi-TeV and may extend to tens of TeVs ranges – probably putting most of them out of range of the current CERN LHC.

Due to the weakness of the ultra-weak interaction, but the anticipated large vacuum expectation values of ρ_1, … , ρ_4, the size of the mass cross terms in the Layer group mass matrices of the different layers is problematic. The mass cross terms (mixing) appear likely to be small.

4.7 The Three Generation Alternative

This chapter and the previous chapter developed a formalism based on four generations of fermions, which necessitated a U(4) Generation group and a U(4) Layer group. If there are only three generations of fermions (as presently observed experimentally) then a U(3) Generation group and a U(3) Layer group would follow.

We continue to believe four generations of fermions will be found and thus the U(4) groups are the correct ones.

5. Normal and Dark Matter Dynamics

While fermions are usually identified by name, it will eventually become difficult to use naming conventions when the 192 fermions in four layers are eventually found. We shall introduce a new naming (identification) convention that uniquely specifies individual fermions in the Periodic Table (front cover).

We shall specify a fermion by a triplet of numbers: species s, layer l, and generation g. The identifying triplet is thus (s, l, g).

Species are numbered from 1 through 12: charged lepton, neutral lepton, three up-type quarks, three down-type quarks, Dark charged lepton, Dark neutral lepton, one Dark up-type quark species, and one Dark down-type quark species.

Layers are numbered from 1 through 4 with our layer being layer 1.

Generations are numbered from 1 through 4 from lightest to the heaviest.[122] (e, ν_e, u and d constitute the known part of generation 1.)

For example, the three b quarks are specified by the triplets (6, 1, 2), (7, 1, 2), and (8, 1, 2). The Dark neutral electron neutrino-like fermion of layer 2 corresponds to the triplet (10, 2, 1).

We now turn to consider the ElectroWeak, Generation group and Layer group interactions. Due to the mixing described in prior chapters of generations due to the ElectroWeak and Generation group Higgs Mechanism the interactions below are "dirty" in the sense that they occur between physical fermion and gauge boson particles that intermix generations. The Layer group also intermixes fermion particles between layers to a very small degree due to its ultra-weak interaction – further muddying the interactions. Thus the below diagrams have *very* small admixtures of other states that are not displayed. Fig. 3.1 in section 3.9 shows the Generation and Layer groups fermion mixing.

END OF EXTRACT FROM Blaha (2016c)

11.4.1 Form of the Layer group Interaction

The U(4) Layer group generators are denoted G_{Li} and its gauge fields are denoted $V_{\mu i}(X)$. Thus the Generation group term in covariant derivatives is

$$g_V \mathbf{V}_\mu \cdot \mathbf{G}_L \tag{11.52}$$

using the 16 U(4) Layer gauge fields $V^a{}_\mu$ with a = 1, 2, …, 16 and the 16 U(4) generators $G_L{}^a$ that couple within, and between layers, using the new layer index where g_v is an ultra-weak coupling constant much smaller than the ElectroWeak coupling constants.

[122] The ordering by mass may not hold in the currently Dark part of the fermion spectrum.

12. A Major Extension of The Standard Model Based on the Riemann-Christoffel Curvature Tensor

This chapter describes the generalization of the Riemann-Christoffel curvature tensor from four Standard Model interactions *to include the three interactions* described in chapter 11.

12.1 Three Additional Interactions for the Extended Standard Model

In chapter 11 we described three 'new' interactions that we have proposed in earlier books in 2015 and 2016 (and before). These interactions were not the result of speculations on symmetry groups beyond those of The Standard Model. They were the result of direct extrapolation of 'known' features of elementary particles:

1. The introduction of the set of complex General Relativistic transformations was prompted by the possibility (reality!) of faster-than-light velocities of neutrinos and down-type quarks which necessitated the transition from the real Lorentz group to the Complex Lorentz group. In turn, the Complex Lorentz group necessarily generalizes to the set of complex General Coordinate transformation – Complex General Relativity. Complex General Relativistic transformations factor into a complex transformation and a real-valued conventional General Relativistic transformation. In chapter 11 we showed that the set of complex transformations is subdivided into the set of those that satisfy the Lorentz condition (eq. 11.50) and the set of those that do not satisfy the condition. See subsection 11.2.10 for details.The second set has a U(4) symmetry, and associated affine connections in curved space and gauge fields in flat space-time. We thus find a need for the change of the term

$$H^{\sigma}{}_{\nu\mu} = \Gamma^{\sigma}{}_{\nu\mu} + \Gamma^{2\sigma}{}_{\nu\mu} \tag{4.53}$$

in the curved space covariant derivative to

$$H^{\sigma}{}_{\nu\mu} = \Gamma_{GR}{}^{\sigma}{}_{\nu\mu} + \Gamma_{GR}{}^{2\sigma}{}_{\nu\mu} + \Gamma_{R}{}^{1\sigma}{}_{\nu\mu} + \Gamma_{R}{}^{2\sigma}{}_{\nu\mu} \tag{11.47}$$

and in flat space-time $\Gamma_R{}^{i\sigma}{}_{\mu\lambda}$ can be expressed in terms of

$$A_{Rflat}{}^{i}{}_{\mu}{}^{a}{}_{b} = A_{Rflat}{}^{i}{}_{\mu}[\tau_k]^{\sigma}{}_{\lambda}\delta_{\sigma}{}^{a}\delta^{\lambda}{}_{b} \tag{11.48}$$

with $A_{Rflat}{}^{ia}{}_{\mu b}$ written in matrix form as

$$A_{Rflat}{}^{i}{}_{\mu} = -\tfrac{1}{2}i\sum_k A_{Rflat}{}^{i}{}_{k\mu}\tau_k \qquad (11.49)$$

2. The apparent conservation of baryon numbers and lepton numbers lead to a broken U(4) symmetry that we call Generation group symmetry since it leads to fermion generations. The U(4) Generation group generators are denoted G_i and its gauge fields are denoted $U_{\mu i}(X)$. Thus the Generation group term in covariant derivatives is

$$g_G \mathbf{U}_\mu \cdot \mathbf{G} \qquad (11.51)$$

with coupling constant g_G.

3. The apparent partial conservation of layer numbers (as described in chapter 11) L_i where i = 1, 2, 3 , 4 leads to the broken U(4) Layer group. The U(4) Layer group generators are denoted G_{Li} and its gauge fields are denoted $V_{\mu i}(X)$. Thus the Generation group term in covariant derivatives is

$$g_V \mathbf{V}_\mu \cdot \mathbf{G}_L \qquad (11.52)$$

using the 16 U(4) Layer gauge fields $V^a{}_\mu$ with layer index number a = 1, 2, …, 16 and 16 U(4) generators $G_L{}^a$ where g_v is the ultra-weak Layer coupling constant.

12.2 Extension of Covariant Derivative

We begin by defining a space vector which also is a fundamental representation vector of each of the factors in SU(3)⊗SU(2)⊗U(1)⊗U(4)⊗U(4)⊗U(4)

$$V_\sigma = V_\sigma{}^{aijkm}(x)\gamma_a T_i \tau_j G_k G_{Lm} \qquad (12.1)$$

Then we use the following generalized covariant derivative of this vector:[123,124]

$$\begin{aligned} D_\nu V_\mu &= (\partial_\nu + iF_\nu)V_\mu - H^\sigma{}_{\nu\mu}V_\sigma \qquad (12.2)\\ &= [g^\sigma{}_\mu\partial_\nu + ig^\sigma{}_\mu F_\nu - H^\sigma{}_{\nu\mu}]V_\sigma \\ &= [g^\sigma{}_\mu\partial_\nu + iD^\sigma{}_{\mu\nu}]V_\sigma \end{aligned}$$

where[125]

$$F_\mu = B^1{}_\mu + A_E{}^1{}_\mu + W^1{}_\mu + A^1{}_\mu + B^2{}_\mu + A_E{}^2{}_\mu + W^2{}_\mu + A^2{}_\mu + \mathbf{U}_\mu + \mathbf{V}_\mu \qquad (12.3)$$

[123] We use the superscript '1' to distinguish primary connections from secondary connections labeled '2'. The discussion in this subsection and in the following subsection parallels that of chapter 4.
[124] Commutator 'cross products' are usually implicit in the following equations.
[125] We will *temporarily* omit the insertion of coupling constants in the interests of simplifying the expressions. We insert them later when we consider the lagrangian.

$$H^{\sigma}{}_{\nu\mu} = \Gamma_{GR}{}^{\sigma}{}_{\nu\mu} + \Gamma_{GR}{}^{2\sigma}{}_{\nu\mu} + \Gamma_{R}{}^{1\sigma}{}_{\nu\mu} + \Gamma_{R}{}^{2\sigma}{}_{\nu\mu} \tag{12.4}$$

$$D^{\sigma}{}_{\mu\nu} = g^{\sigma}{}_{\mu}F_{\nu} + iH^{\sigma}{}_{\nu\mu} \tag{12.5}$$

Commutators of the vector fields in F_{μ} are implicit when the covariant derivative is applied to vectors and tensors such as V_{σ} in eq. 12.1.

Eq. 12.2 enables us to derive the Riemann-Christoffel curvature tensor, and then its contractions $R_{\mu\nu}$ and R using

$$(D_{\nu}D_{\mu} - D_{\mu}D_{\nu})V_{\sigma} = R^{\beta}{}_{\sigma\nu\mu}V_{\beta} \tag{12.6}$$

The second order covariant derivative of V_{σ} is

$$D_{\nu}D_{\mu}V_{\sigma} = \{g^{\alpha}{}_{\mu}(\partial_{\nu} + iF_{\nu}) - H^{\alpha}{}_{\mu\nu}\}\{g^{\beta}{}_{\sigma}(\partial_{\alpha} + iF_{\alpha})V_{\beta} - H^{\beta}{}_{\sigma\alpha}V_{\beta}\} - H^{\gamma}{}_{\nu\sigma}\{g^{\alpha}{}_{\gamma}(\partial_{\mu} + iF_{\mu})V_{\alpha} - H^{\alpha}{}_{\gamma\mu}V_{\alpha}\} \tag{12.7}$$

12.3 Extension of Riemann-Christoffel Curvature Tensor

In chapter 4 we calculated the Riemann-Christoffel curvature tensor $R'^{\beta}{}_{\sigma\nu\mu}$ for General Relativity and the Standard Model SU(3)⊗SU(2)⊗U(1) symmetry. In this chapter we extend the calculation to include the three additional interactions determined in chapter 11 and specified above in section 12.1 for the extension of The Standard Model to SU(3)⊗SU(2)⊗U(1)⊗U(4)⊗U(4)⊗U(4). The calculation is similar to that given in chapter 4.

We find

$$\begin{aligned} R'^{\beta}{}_{\sigma\nu\mu}V_{\beta} &= g^{\alpha}{}_{\mu}(\partial_{\nu} + iF_{\nu})g^{\beta}{}_{\sigma}(\partial_{\alpha} + iF_{\alpha})V_{\beta} - H^{\alpha}{}_{\mu\nu}g^{\beta}{}_{\sigma}(\partial_{\alpha} + iF_{\alpha})V_{\beta} + \\ &\quad + H^{\alpha}{}_{\mu\nu}H^{\beta}{}_{\sigma\alpha}V_{\beta} - g^{\alpha}{}_{\mu}(\partial_{\nu} + iF_{\nu})H^{\beta}{}_{\sigma\alpha}V_{\beta} - H^{\gamma}{}_{\nu\sigma}\{g^{\alpha}{}_{\gamma}(\partial_{\mu} + iF_{\mu})V_{\alpha} - H^{\alpha}{}_{\gamma\mu}V_{\alpha}\} - \\ &\quad - \{\mu\leftrightarrow\nu\} \\ &= ig^{\beta}{}_{\sigma}(\partial_{\nu}F_{\mu} - \partial_{\mu}F_{\nu} - i[F_{\nu}, F_{\mu}])V_{\beta} + (\partial_{\mu}H^{\beta}{}_{\sigma\nu} - \partial_{\nu}H^{\beta}{}_{\sigma\mu} + H^{\gamma}{}_{\nu\sigma}H^{\beta}{}_{\gamma\mu} - H^{\gamma}{}_{\mu\sigma}H^{\beta}{}_{\gamma\nu})V_{\beta} \\ &= ig^{\beta}{}_{\sigma}(F_{E}{}^{1}{}_{\nu\mu} + F_{E}{}^{2}{}_{\nu\mu} + F_{W}{}^{1}{}_{\nu\mu} + F_{W}{}^{2}{}_{\nu\mu} + F^{1}{}_{\nu\mu} + F^{2}{}_{\nu\mu} + F_{U}{}^{1}{}_{\nu\mu} + F_{U}{}^{2}{}_{\nu\mu} + F_{V}{}^{1}{}_{\nu\mu} + F_{V}{}^{2}{}_{\nu\mu})V_{\beta} + \\ &\quad + (ig^{\beta}{}_{\sigma}B^{1}{}_{\nu\mu} + ig^{\beta}{}_{\sigma}B^{2}{}_{\nu\mu} + \partial_{\mu}H^{\beta}{}_{\sigma\nu} - \partial_{\nu}H^{\beta}{}_{\sigma\mu} + H^{\gamma}{}_{\nu\sigma}H^{\beta}{}_{\gamma\mu} - H^{\gamma}{}_{\mu\sigma}H^{\beta}{}_{\gamma\nu})V_{\beta} \\ &= R'_{E}{}^{\beta}{}_{\sigma\nu\mu}V_{\beta} + R'_{SU(2)}{}^{\beta}{}_{\sigma\nu\mu}V_{\beta} + R'_{SU(3)}{}^{\beta}{}_{\sigma\nu\mu}V_{\beta} + R'_{U}{}^{\beta}{}_{\sigma\nu\mu}V_{\beta} + R'_{V}{}^{\beta}{}_{\sigma\nu\mu}V_{\beta} + R'_{G}{}^{\beta}{}_{\sigma\nu\mu}V_{\beta} \end{aligned} \tag{12.8}$$

where

$$\begin{aligned} R'_{SU(3)}{}^{\beta}{}_{\sigma\nu\mu} &= ig^{\beta}{}_{\sigma}(F^{1}{}_{\nu\mu} + F^{2}{}_{\nu\mu}) \\ R'_{SU(2)}{}^{\beta}{}_{\sigma\nu\mu} &= ig^{\beta}{}_{\sigma}(F_{W}{}^{1}{}_{\nu\mu} + F_{W}{}^{2}{}_{\nu\mu}) \\ R'_{E}{}^{\beta}{}_{\sigma\nu\mu} &= ig^{\beta}{}_{\sigma}(F_{E}{}^{1}{}_{\nu\mu} + F_{E}{}^{2}{}_{\nu\mu}) \\ R'_{U}{}^{\beta}{}_{\sigma\nu\mu} &= ig^{\beta}{}_{\sigma}(F_{U}{}^{1}{}_{\nu\mu} + F_{U}{}^{2}{}_{\nu\mu}) \\ R'_{V}{}^{\beta}{}_{\sigma\nu\mu} &= ig^{\beta}{}_{\sigma}(F_{V}{}^{1}{}_{\nu\mu} + F_{V}{}^{2}{}_{\nu\mu}) \end{aligned} \tag{12.9}$$

and

$$R'_G{}^{\beta}{}_{\sigma\nu\mu} = ig^{\beta}{}_{\sigma}(B^1{}_{\nu\mu} + B^2{}_{\nu\mu}) + \partial_\mu H^{1\beta}{}_{\sigma\nu} - \partial_\nu H^{1\beta}{}_{\sigma\mu} + H^{1\gamma}{}_{\nu\sigma}H^{1\beta}{}_{\gamma\mu} - H^{1\gamma}{}_{\mu\sigma}H^{1\beta}{}_{\gamma\nu} + \partial_\mu H^{2\beta}{}_{\sigma\nu} - \partial_\nu H^{2\beta}{}_{\sigma\mu} +$$
$$+ H^{2\gamma}{}_{\nu\sigma}H^{2\beta}{}_{\gamma\mu} - H^{2\gamma}{}_{\mu\sigma}H^{2\beta}{}_{\gamma\nu} + H^{1\gamma}{}_{\nu\sigma}H^{2\beta}{}_{\gamma\mu} - H^{1\gamma}{}_{\mu\sigma}H^{2\beta}{}_{\gamma\nu} + H^{2\gamma}{}_{\nu\sigma}H^{1\beta}{}_{\gamma\mu} - \Gamma^{2\gamma}{}_{\mu\sigma}\Gamma^{\beta}{}_{\gamma\nu} \quad (12.10)$$
$$= ig^{\beta}{}_{\sigma}(B^1{}_{\nu\mu} + B^2{}_{\nu\mu}) + R^{1\beta}{}_{\sigma\nu\mu} + R^{2\beta}{}_{\sigma\nu\mu}$$

with

$$H^{\beta}{}_{\sigma\nu\mu} = \partial_\mu H^{\beta}{}_{\sigma\nu} - \partial_\nu H^{\beta}{}_{\sigma\mu} + H^{\gamma}{}_{\nu\sigma}H^{\beta}{}_{\gamma\mu} - H^{\gamma}{}_{\mu\sigma}H^{\beta}{}_{\gamma\nu} \quad (12.11)$$
$$R^{1\beta}{}_{\sigma\nu\mu} = \partial_\mu H^{1\beta}{}_{\sigma\nu} - \partial_\nu H^{1\beta}{}_{\sigma\mu} + H^{1\gamma}{}_{\nu\sigma}H^{1\beta}{}_{\gamma\mu} - H^{1\gamma}{}_{\mu\sigma}H^{1\beta}{}_{\gamma\nu} \quad (12.12)$$
$$R^{2\beta}{}_{\sigma\nu\mu\rho} = \partial_\mu H^{2\beta}{}_{\sigma\nu} - \partial_\nu H^{2\beta}{}_{\sigma\mu} + H^{2\gamma}{}_{\nu\sigma}H^{2\beta}{}_{\gamma\mu} - H^{2\gamma}{}_{\mu\sigma}H^{2\beta}{}_{\gamma\nu} +$$
$$+ H^{1\gamma}{}_{\nu\sigma}H^{2\beta}{}_{\gamma\mu} - H^{1\gamma}{}_{\mu\sigma}H^{2\beta}{}_{\gamma\nu} + H^{2\gamma}{}_{\nu\sigma}H^{1\beta}{}_{\gamma\mu} - H^{2\gamma}{}_{\mu\sigma}H^{1\beta}{}_{\gamma\nu} \quad (12.13)$$

and

$$H^{1\sigma}{}_{\nu\mu} = \Gamma_{GR}{}^{\sigma}{}_{\nu\mu} + \Gamma_R{}^{1\sigma}{}_{\nu\mu}$$
$$H^{2\sigma}{}_{\nu\mu} = \Gamma_{GR}{}^{2\sigma}{}_{\nu\mu} + \Gamma_R{}^{2\sigma}{}_{\nu\mu}$$

and where

$$F^1{}_{\kappa\mu} = \partial A^1{}_\mu/\partial x^\kappa - \partial A^1{}_\kappa/\partial x^\mu + i[A^1{}_\kappa, A^1{}_\mu] \quad (12.14)$$
$$F_W{}^1{}_{\kappa\mu} = \partial W^1{}_\mu/\partial x^\kappa - \partial W^1{}_\kappa/\partial x^\mu + i[W^1{}_\kappa, W^1{}_\mu]$$
$$F_E{}^1{}_{\kappa\mu} = \partial A_E{}^1{}_\mu/\partial x^\kappa - \partial A_E{}^1{}_\kappa/\partial x^\mu$$
$$B^1{}_{\kappa\mu} = \partial B^1{}_\mu/\partial x^\kappa - \partial B^1{}_\kappa/\partial x^\mu + i[B^1{}_\kappa, B^1{}_\mu]$$
$$F_U{}^1{}_{\kappa\mu} = \partial U^1{}_\mu/\partial x^\kappa - \partial U^1{}_\kappa/\partial x^\mu + i[U^1{}_\kappa, U^1{}_\mu]$$
$$F_V{}^1{}_{\kappa\mu} = \partial V^1{}_\mu/\partial x^\kappa - \partial V^1{}_\kappa/\partial x^\mu + i[V^1{}_\kappa, V^1{}_\mu]$$

$$F^2{}_{\kappa\mu} = \partial A^2{}_\mu/\partial x^\kappa - \partial A^2{}_\kappa/\partial x^\mu + i[A^2{}_\kappa, A^2{}_\mu] + i[A^1{}_\kappa, A^2{}_\mu] + i[A^2{}_\kappa, A^1{}_\mu]$$
$$F_W{}^2{}_{\kappa\mu} = \partial W^2{}_\mu/\partial x^\kappa - \partial W^2{}_\kappa/\partial x^\mu + i[W^2{}_\kappa, W^2{}_\mu] + i[W^1{}_\kappa, W^2{}_\mu] + i[W^2{}_\kappa, W^1{}_\mu]$$
$$F_E{}^2{}_{\kappa\mu} = \partial A_E{}^2{}_\mu/\partial x^\kappa - \partial A_E{}^2{}_\kappa/\partial x^\mu$$
$$B^2{}_{\kappa\mu} = \partial B^2{}_\mu/\partial x^\kappa - \partial B^2{}_\kappa/\partial x^\mu + i[B^2{}_\mu, B^2{}_\kappa] + i[B^1{}_\mu, B^2{}_\kappa] + i[B^2{}_\mu, B^1{}_\kappa]$$
$$F_U{}^2{}_{\kappa\mu} = \partial U^2{}_\mu/\partial x^\kappa - \partial U^2{}_\kappa/\partial x^\mu + i[U^2{}_\kappa, U^2{}_\mu] + i[U^1{}_\kappa, U^2{}_\mu] + i[U^2{}_\kappa, U^1{}_\mu]$$
$$F_V{}^2{}_{\kappa\mu} = \partial V^2{}_\mu/\partial x^\kappa - \partial V^2{}_\kappa/\partial x^\mu + i[V^2{}_\kappa, V^2{}_\mu] + i[V^1{}_\kappa, V^2{}_\mu] + i[V^2{}_\kappa, V^1{}_\mu]$$

$R''^{\beta}{}_{\sigma\nu\mu}$ *is the Riemann-Christoffel curvature tensor for the complex 32-dimensional space generated by SU(3) and 4-dimensional General Coordinate transformations including secondary connections and metric with U(1)*⊗SU(2)⊗SU(3)⊗U(4)⊗U(4)⊗U(4) *internal symmetry* .

Note that $R''^{\beta}{}_{\sigma\nu\mu}$ factorizes below into U(1)⊗SU(2)⊗SU(3)⊗U(4)⊗U(4) parts and a Riemann-Christoffel Gravitational curvature tensor part with a U(4) part due to the commutativity amongst $B^i{}_\mu$, $A^j{}_\mu$, $W^k{}_\mu$, $A_E{}^k{}_\mu$, $\mathbf{U}^m{}_\mu$ and $\mathbf{V}^n{}_\mu$ for all i, j, k, m, and n.

For later use in defining a lagrangian we define

$$R'^{\beta}{}_{\sigma\nu\mu} = R'^1{}_E{}^{\beta}{}_{\sigma\nu\mu} + R'^2{}_E{}^{\beta}{}_{\sigma\nu\mu} + R'^1{}_{SU(2)}{}^{\beta}{}_{\sigma\nu\mu} + R'^2{}_{SU(2)}{}^{\beta}{}_{\sigma\nu\mu} + R'^1{}_{SU(3)}{}^{\beta}{}_{\sigma\nu\mu} + R'^2{}_{SU(3)}{}^{\beta}{}_{\sigma\nu\mu} +$$
$$+ R'^1{}_U{}^{\beta}{}_{\sigma\nu\mu} + R'^2{}_U{}^{\beta}{}_{\sigma\nu\mu} + R'^1{}_V{}^{\beta}{}_{\sigma\nu\mu} + R'^2{}_V{}^{\beta}{}_{\sigma\nu\mu} +$$

$$+ R'^{1}{}_{B}{}^{\beta}{}_{\sigma\nu\mu} + R'^{2}{}_{B}{}^{\beta}{}_{\sigma\nu\mu} + R^{1\beta}{}_{\sigma\nu\mu} + R^{2\beta}{}_{\sigma\nu\mu} \tag{12.15}$$

where

$$R'^{1}{}_{E}{}^{\beta}{}_{\sigma\nu\mu} = ig^{\beta}{}_{\sigma}F_{E}{}^{1}{}_{\nu\mu} \tag{12.16}$$
$$R'^{2}{}_{E}{}^{\beta}{}_{\sigma\nu\mu} = ig^{\beta}{}_{\sigma}F_{E}{}^{2}{}_{\nu\mu}$$
$$R'^{1}{}_{SU(2)}{}^{\beta}{}_{\sigma\nu\mu} = ig^{\beta}{}_{\sigma}F_{W}{}^{1}{}_{\nu\mu} \tag{12.17}$$
$$R'^{2}{}_{SU(2)}{}^{\beta}{}_{\sigma\nu\mu} = ig^{\beta}{}_{\sigma}F_{W}{}^{2}{}_{\nu\mu}$$

$$R'^{1}{}_{SU(3)}{}^{\beta}{}_{\sigma\nu\mu} = ig^{\beta}{}_{\sigma}F^{1}{}_{\nu\mu} \tag{12.18}$$
$$R'^{2}{}_{SU(3)}{}^{\beta}{}_{\sigma\nu\mu} = ig^{\beta}{}_{\sigma}F^{2}{}_{\nu\mu}$$

$$R'^{1}{}_{U}{}^{\beta}{}_{\sigma\nu\mu} = ig^{\beta}{}_{\sigma}F_{U}{}^{1}{}_{\nu\mu} \tag{12.18u}$$
$$R'^{2}{}_{U}{}^{\beta}{}_{\sigma\nu\mu} = ig^{\beta}{}_{\sigma}F_{U}{}^{2}{}_{\nu\mu}$$

$$R'^{1}{}_{V}{}^{\beta}{}_{\sigma\nu\mu} = ig^{\beta}{}_{\sigma}F_{V}{}^{1}{}_{\nu\mu} \tag{12.18v}$$
$$R'^{2}{}_{V}{}^{\beta}{}_{\sigma\nu\mu} = ig^{\beta}{}_{\sigma}F_{V}{}^{2}{}_{\nu\mu}$$

$$R'^{1}{}_{B}{}^{\beta}{}_{\sigma\nu\mu} = ig^{\beta}{}_{\sigma}B^{1}{}_{\nu\mu} \tag{12.19}$$
$$R'^{2}{}_{B}{}^{\beta}{}_{\sigma\nu\mu} = ig^{\beta}{}_{\sigma}B^{2}{}_{\nu\mu}$$

The total Ricci tensor is

$$R'_{\sigma\mu} = R'^{\beta}{}_{\sigma\beta\mu} \tag{12.20}$$

$$= iF_{E}{}^{1}{}_{\sigma\mu} + iF_{E}{}^{2}{}_{\sigma\mu} + iF_{W}{}^{1}{}_{\sigma\mu} + iF_{W}{}^{2}{}_{\sigma\mu} + iF^{1}{}_{\sigma\mu} + iF^{2}{}_{\sigma\mu} + iF_{U}{}^{1}{}_{\sigma\mu} + iF_{U}{}^{2}{}_{\sigma\mu} + iF_{V}{}^{1}{}_{\sigma\mu} + iF_{V}{}^{2}{}_{\sigma\mu} + iB^{1}{}_{\sigma\mu} + iB^{2}{}_{\sigma\mu} +$$
$$+ \partial_{\mu}H^{1\beta}{}_{\sigma\beta} - \partial_{\beta}H^{1\beta}{}_{\sigma\mu} + H^{1\gamma}{}_{\beta\sigma}H^{1\beta}{}_{\gamma\mu} - H^{1\gamma}{}_{\mu\sigma}H^{1\beta}{}_{\gamma\beta} +$$
$$+ \partial_{\mu}H^{2\beta}{}_{\sigma\beta} - \partial_{\beta}H^{2\beta}{}_{\sigma\mu} + H^{2\gamma}{}_{\beta\sigma}H^{2\beta}{}_{\gamma\mu} - H^{2\gamma}{}_{\mu\sigma}H^{2\beta}{}_{\gamma\beta} + H^{1\gamma}{}_{\beta\sigma}H^{2\beta}{}_{\gamma\mu} - H^{1\gamma}{}_{\mu\sigma}H^{2\beta}{}_{\gamma\beta} + H^{2\gamma}{}_{\beta\sigma}H^{1\beta}{}_{\gamma\mu} - H^{2\gamma}{}_{\mu\sigma}H^{1\beta}{}_{\gamma\beta}$$

$$= R'^{1}{}_{E}{}^{\beta}{}_{\sigma\beta\mu} + R'^{2}{}_{E}{}^{\beta}{}_{\sigma\beta\mu} + R'^{1}{}_{SU(2)}{}^{\beta}{}_{\sigma\beta\mu} + R'^{2}{}_{SU(2)}{}^{\beta}{}_{\sigma\beta\mu} + R'^{1}{}_{SU(3)}{}^{\beta}{}_{\sigma\beta\mu} + R'^{2}{}_{SU(3)}{}^{\beta}{}_{\sigma\beta\mu} + R'^{1}{}_{U}{}^{\beta}{}_{\sigma\beta\mu} + R'^{2}{}_{U}{}^{\beta}{}_{\sigma\beta\mu} +$$
$$+ R'^{1}{}_{V}{}^{\beta}{}_{\sigma\beta\mu} + R'^{2}{}_{V}{}^{\beta}{}_{\sigma\beta\mu} + R'^{1}{}_{B}{}^{\beta}{}_{\sigma\beta\mu} + R'^{2}{}_{B}{}^{\beta}{}_{\sigma\beta\mu} + R^{1\beta}{}_{\sigma\beta\mu} + R^{2\beta}{}_{\sigma\beta\mu}$$
$$= R'^{1}{}_{E\sigma\mu} + R'^{2}{}_{E\sigma\mu} + R'^{1}{}_{SU(2)\sigma\mu} + R'^{2}{}_{SU(2)\sigma\mu} + R'^{1}{}_{SU(3)\sigma\mu} + R'^{2}{}_{SU(3)\sigma\mu} + R'^{1}{}_{U\sigma\mu} + R'^{2}{}_{U\sigma\mu} + R'^{1}{}_{V\sigma\mu} +$$
$$+ R'^{2}{}_{V\sigma\mu} + R'^{1}{}_{B}{}^{\beta}{}_{\sigma\beta\mu} + R'^{2}{}_{B}{}^{\beta}{}_{\sigma\beta\mu} + R^{1}{}_{\sigma\mu} + R^{2}{}_{\sigma\mu}$$
$$= R'^{1}{}_{\sigma\mu} + R'^{2}{}_{\sigma\mu} \tag{12.21}$$

where

$$R'^{1}{}_{\sigma\mu} = R'^{1}{}_{E\sigma\mu} + R'^{1}{}_{SU(2)\sigma\mu} + R'^{1}{}_{SU(3)\sigma\mu} + R'^{1}{}_{U\sigma\mu} + R'^{1}{}_{V\sigma\mu} + R'^{1}{}_{B}{}^{\beta}{}_{\sigma\beta\mu} + R^{1}{}_{\sigma\mu} \tag{12.22}$$
$$R'^{2}{}_{\sigma\mu} = R'^{2}{}_{E\sigma\mu} + R'^{2}{}_{SU(2)\sigma\mu} + R'^{2}{}_{SU(3)\sigma\mu} + R'^{2}{}_{U\sigma\mu} + R'^{2}{}_{V\sigma\mu} + R'^{2}{}_{B}{}^{\beta}{}_{\sigma\beta\mu} + R^{2}{}_{\sigma\mu}$$
$$R'^{1}{}_{E\sigma\mu} = iF_{E}{}^{1}{}_{\sigma\mu}$$
$$R'^{2}{}_{E\sigma\mu} = iF_{E}{}^{2}{}_{\sigma\mu}$$

$$R'^{1}{}_{SU(2)\sigma\mu} = iF_{W}{}^{1}{}_{\sigma\mu}$$
$$R'^{2}{}_{SU(2)\sigma\mu} = iF_{W}{}^{2}{}_{\sigma\mu}$$

$$R'^{1}{}_{SU(3)\sigma\mu} = iF^{1}{}_{\sigma\mu}$$
$$R'^{2}{}_{SU(3)\sigma\mu} = iF^{2}{}_{\sigma\mu}$$

$$R'^{1}{}_{U\sigma\mu} = iF_{U}{}^{1}{}_{\sigma\mu}$$
$$R'^{2}{}_{U\sigma\mu} = iF_{U}{}^{2}{}_{\sigma\mu}$$

$$R'^{1}{}_{V\sigma\mu} = iF_{V}{}^{1}{}_{\sigma\mu}$$
$$R'^{2}{}_{V\sigma\mu} = iF_{V}{}^{2}{}_{\sigma\mu}$$

$$R'^{1}{}_{B\sigma\mu} = iB^{1}{}_{\sigma\mu}$$
$$R'^{2}{}_{B\sigma\mu} = iB^{2}{}_{\sigma\mu}$$

with the further definition of $R''^{1}{}_{\sigma\mu}$ and $R''^{2}{}_{\sigma\mu}$:[126]

$$R''^{1}{}_{\sigma\mu} = R'^{1}{}_{\sigma\mu} - R'^{1}{}_{E\sigma\mu} - R'^{1}{}_{W\sigma\mu} - R'^{1}{}_{U\sigma\mu} - R'^{1}{}_{V\sigma\mu} \qquad (12.23)$$
$$R''^{2}{}_{\sigma\mu} = R'^{2}{}_{\sigma\mu} - R'^{2}{}_{E\sigma\mu} - R'^{2}{}_{W\sigma\mu} - R'^{2}{}_{U\sigma\mu} - R'^{2}{}_{V\sigma\mu}$$

Eq. 12.21 is the Ricci tensor for the complex 32-dimensional space generated by SU(3) and 4-dimensional General Coordinate transformations when augmented with the secondary affine connections.

An additional Ricci-like tensor is

$$H_{\sigma\mu} = H^{\beta}{}_{\sigma\beta\mu} \qquad (12.24)$$

The curvature scalar is

$$R' = g^{\sigma\mu}R'_{\sigma\mu} = + \partial^{\sigma}H^{1\beta}{}_{\sigma\beta} - \partial_{\beta}H^{1\beta}{}_{\sigma}{}^{\sigma} + H^{1\gamma}{}_{\beta\sigma}H^{1\beta}{}_{\gamma}{}^{\sigma} - H^{1\gamma}{}_{\mu\sigma}H^{1\beta}{}_{\gamma\beta} + \partial^{\sigma}H^{2\beta}{}_{\sigma\beta} - \partial_{\beta}H^{2\beta}{}_{\sigma}{}^{\sigma} +$$
$$+ H^{2\gamma}{}_{\beta\sigma}H^{2\beta}{}_{\gamma}{}^{\sigma} - H^{2\gamma\sigma}{}_{\sigma}H^{2\beta}{}_{\gamma\beta} + H^{1\gamma}{}_{\beta\sigma}H^{2\beta}{}_{\gamma}{}^{\sigma} - H^{1\gamma\sigma}{}_{\sigma}H^{2\beta}{}_{\gamma\beta} + H^{2\gamma}{}_{\beta\sigma}H^{1\beta}{}_{\gamma}{}^{\sigma} - H^{2\gamma\sigma}{}_{\sigma}H^{1\beta}{}_{\gamma\beta}$$

$$= g^{\sigma\mu}(R^{1\beta}{}_{\sigma\beta\mu} + R^{2\beta}{}_{\sigma\beta\mu}) \qquad (12.25)$$

Additional curvature scalars are

$$H = g^{\sigma\mu}H_{\sigma\mu} \qquad (12.26)$$
$$R'^{2} = g^{\sigma\mu}R'^{2}{}_{\sigma\mu} \qquad (12.27)$$

---

[126] We subtract the $R'^{i}{}_{U\sigma\mu}$ and $R'^{i}{}_{V\sigma\mu}$ terms for i = 1, 2 from $R''^{i}{}_{\sigma\mu}$ since we view U and V symmetry as having the same 'nature' as the Weak interaction symmetry. If *not* subtracted the U and V gauge bosons would acquire masses from the lagrangian below (eq. 12.28).

12.4 Total Vector Boson and Gravitational Lagrangian

In this section we will define a total Vector boson and gravitational lagrangian as a generalization of the usual Einstein lagrangian with additional higher derivative terms added.

The major aspect of our extension is the introduction of higher derivative terms in a such a manner that they can be handled by canonical lagrangian methods to obtain the dynamical equations of motion and the equal time commutation relations (for the 'free' field approximations.)

We also separate the Ricci tensor into two parts in order to use pseudoQuantization field theory (with fields labeled '1' and '2') to implement canonical lagrangian methods, and to introduce the flat space metric $\eta^{\sigma\mu}$ by a Higgs Mechanism.[127] The constant flat space metric part $\eta^{\sigma\mu}$ of the weak field quantum metric is usually an assumed quantity. But its close relation to the quantum field $g^{\sigma\mu}$ suggests that it could be generated by the same Higgs Mechanism that generates particle mass constants.[128]

We assume the lagrangian density:[129]

$$\mathcal{L} = \text{Tr}\ \sqrt{g}[MD_\nu R''^1{}_{\sigma\mu}D^\nu R''^{2\sigma\mu} + aR'^1{}_{\sigma\mu}R'^{2\sigma\mu} + bR' + cg^{\sigma\mu}g^2{}_{\sigma\mu} + c'g^{2\sigma\mu}g^2{}_{\sigma\mu} - dA^2{}_\mu A^{2\mu}] \qquad (12.28)$$

where M, a, b, c, c', and d are constants to be determined later.

This higher derivative lagrangian maintains the locality of the theory but does entail a modest modification in the derivation of the Euler-Lagrange equations of motion. It also requires the use of principal value propagators rather than ordinary Feynman propagators for gluon and graviton interactions. Thus the Strong Interaction sector, and the Gravitation sector are Action-at-a-Distance theories that are similar in spirit to Wheeler-Feynman Electrodynamics. The Electromagnetic sector, the U sector, the V sector, and Weak sector may, or may not, be an Action-at-a-Distance theory. They are not constrained to be Action-at-a-Distance by the present considerations.

Since we wish to apply it cosmologically, and within hadrons, where the gravitational spinor connections are negligible due to the smallness of the gravitational constant G and the 'smallness' of B spin on the cosmological scale, we set $B^1{}_{\nu\mu} = B^2{}_{\nu\mu} = 0$ and find[130]

$$\begin{aligned}\mathcal{L} = \text{Tr}\ \sqrt{g}[&MD_\nu(R'^1{}_{SU(3)\sigma\mu} + R'^1{}_{G\sigma\mu})D^\nu(R'^2{}_{SU(3)}{}^{\sigma\mu} + R'^2{}_G{}^{\sigma\mu}) + \\ &+ a(R'^1{}_{E\sigma\mu} + R'^1{}_{SU(2)\sigma\mu} + R'^1{}_{SU(3)\sigma\mu} + R'^1{}_{U\sigma\mu} + R'^1{}_{V\sigma\mu} + R^1{}_{\sigma\mu})\cdot \\ &\cdot(R'^2{}_E{}^{\sigma\mu} + R'^2{}_{SU(2)}{}^{\sigma\mu} + R'^2{}_{SU(3)}{}^{\sigma\mu} + R'^2{}_U{}^{\sigma\mu} + R'^2{}_V{}^{\sigma\mu} + R^{2\sigma\mu}) +\end{aligned}$$

---

[127] In earlier books such as Blaha (2016f) we showed that the use of two fields for each particle type enables us to clearly separate the 'vacuum expectation value' from its associated second quantized 'Higgs' field. The application to the weak field approximation for gravitons is one example.

[128] See also Blaha (2016c).

[129] Since the lagrangian terms are matrices it is necessary to take the trace.

[130] The constants in eq. 4.62 have the dimensions: M has the dimension of inverse mass squared, b has dimension mass squared, a is dimensionless, c and c' have dimension mass to the 4th order, and d has dimension mass squared.

$$+ bR' + cg^{\sigma\mu}g^2{}_{\sigma\mu} + c'g^{2\sigma\mu}g^2{}_{\sigma\mu} - dA^2{}_\mu A^{2\mu}]$$

$$\begin{aligned}
&= \mathrm{Tr}\,\sqrt{g}[M\{D_\nu R'^1{}_{SU(3)\sigma\mu}D^\nu R'^2{}_{SU(3)}{}^{\sigma\mu} + D_\nu R^1{}_{\sigma\mu}D^\nu R^{2\sigma\mu} + D_\nu R^1{}_{\sigma\mu}D^\nu R'^2{}_{SU(3)}{}^{\sigma\mu} + D_\nu R'^1{}_{SU(3)\sigma\mu}D^\nu R^{2\sigma\mu}\} + \\
&\quad + a\{(R'^1{}_{E\sigma\mu} + R'^1{}_{SU(2)\sigma\mu} + R'^1{}_{SU(3)\sigma\mu} + R'^1{}_{U\sigma\mu} + R'^1{}_{V\sigma\mu})\cdot \\
&\qquad \cdot(R'^2{}_E{}^{\sigma\mu} + R'^2{}_{SU(2)}{}^{\sigma\mu} + R'^2{}_{SU(3)}{}^{\sigma\mu} + R'^2{}_U{}^{\sigma\mu} + R'^2{}_V{}^{\sigma\mu}) + \\
&\quad + R^1{}_{\sigma\mu}R^{2\sigma\mu} + R^1{}_{\sigma\mu}(R'^2{}_E{}^{\sigma\mu} + R'^2{}_{SU(2)}{}^{\sigma\mu} + R'^2{}_{SU(3)}{}^{\sigma\mu} + R'^2{}_U{}^{\sigma\mu} + R'^2{}_V{}^{\sigma\mu}) + \\
&\quad + (R'^1{}_{E\sigma\mu} + R'^1{}_{SU(2)\sigma\mu} + R'^1{}_{SU(3)\sigma\mu} + R'^1{}_{U\sigma\mu} + R'^1{}_{V\sigma\mu})R^{2\sigma\mu}\} + \\
&\quad + bg^{\sigma\mu}(R^{1\beta}{}_{\sigma\beta\mu} + R^{2\beta}{}_{\sigma\beta\mu}) + cg^{\sigma\mu}g^2{}_{\sigma\mu} + c'g^{2\sigma\mu}g^2{}_{\sigma\mu} - dA^2{}_\mu A^{2\mu}] \qquad (12.29)
\end{aligned}$$

Since there are no strong interaction fields in 'empty' space and gravity is negligible within hadrons,[131] we can drop the interaction terms between these interactions. However, we cannot drop the interaction terms amongst Electromagnetism, the Weak interaction, the Strong Interaction, the Generation group U interaction and the Layer group V interaction, within, and between, hadrons, and the interaction terms between Electromagnetism and Gravitation cosmologically.

Eq. 4.79 can therefore be expressed as:[132]

$$\mathcal{L} = \mathcal{L}_E + \mathcal{L}_{SU(2)} + \mathcal{L}_{SU(3)} + \mathcal{L}_U + \mathcal{L}_V + \mathcal{L}_G + \mathcal{L}_{int} \qquad (12.30)$$

where taking traces of $\mathcal{L}$s terms is understood[133]

$$\begin{aligned}
\mathcal{L}_E &= \mathrm{Tr}\,\sqrt{g}\{M\{[\partial_\nu + i(A_E{}^1{}_\nu + A_E{}^2{}_\nu)]F^1{}_{E\sigma\mu}[\partial^\nu + i(A_E{}^{1\nu} + A_E{}^{2\nu})]F^2{}_E{}^{\sigma\mu}\} + aF^1{}_{E\sigma\mu}F^2{}_E{}^{\sigma\mu}\} \qquad (12.31) \\
\mathcal{L}_{SU(2)} &= \mathrm{Tr}\,\sqrt{g}[aF_W{}^1{}_{\sigma\mu}F_W{}^{2\sigma\mu}] \\
\mathcal{L}_{SU(3)} &= \mathrm{Tr}\,\sqrt{g}\{M[\partial_\nu + i(A^1{}_\nu + A^2{}_\nu)]F^1{}_{\sigma\mu}[\partial^\nu + i(A^{1\nu} + A^{2\nu})]F^{2\sigma\mu} + aF^1{}_{\sigma\mu}F^{2\sigma\mu} - dA^2{}_\mu A^{2\mu}\} \\
\mathcal{L}_U &= \mathrm{Tr}\,\sqrt{g}[aF_U{}^1{}_{\sigma\mu}F_U{}^{2\sigma\mu}] \\
\mathcal{L}_V &= \mathrm{Tr}\,\sqrt{g}[aF_V{}^1{}_{\sigma\mu}F_V{}^{2\sigma\mu}]
\end{aligned}$$

$$\begin{aligned}
\mathcal{L}_G &= \mathrm{Tr}\,\sqrt{g}[MD_\nu R^1{}_{\sigma\mu}D^\nu R^{2\sigma\mu} + aR^1{}_{\sigma\mu}R^{2\sigma\mu} + bg^{\sigma\mu}(R^{1\beta}{}_{\sigma\beta\mu} + R^{2\beta}{}_{\sigma\beta\mu}) + cg^{\sigma\mu}g^2{}_{\sigma\mu} + c'g^{2\sigma\mu}g^2{}_{\sigma\mu}] \\
&= \mathrm{Tr}\,\sqrt{g}[MD_\nu R^1{}_{\sigma\mu}D^\nu R^{2\sigma\mu} + aR^1{}_{\sigma\mu}R^{2\sigma\mu} + bH + cg^{\sigma\mu}g^2{}_{\sigma\mu} + c'g^{2\sigma\mu}g^2{}_{\sigma\mu}]
\end{aligned}$$

$$\begin{aligned}
\mathcal{L}_{int} = \mathrm{Tr}\,\sqrt{g}[M\{&-i(A_E{}^1{}_\nu + A_E{}^2{}_\nu + W^1{}_\nu + W^2{}_\nu + U^1{}_\nu + U^2{}_\nu + V^1{}_\nu + V^2{}_\nu)F^1{}_{\sigma\mu}D^\nu F^{2\sigma\mu} - \\
&- iD_\nu F^1{}_{\sigma\mu}(A_E{}^{1\nu} + A_E{}^{2\nu} + W^{1\nu} + W^{2\nu} + U^{1\nu} + U^{2\nu} + V^{1\nu} + V^{2\nu})F^{2\sigma\mu} + \\
&+ iD_\nu R^1{}_{\sigma\mu}[\partial^\nu + i(A^{1\nu} + A^{2\nu} + A_E{}^{1\nu} + A_E{}^{2\nu} + W^{1\nu} + W^{2\nu} + U^{1\nu} + U^{2\nu} + V^{1\nu} + V^{2\nu})]F^{2\sigma\mu} + \\
&+ i[\partial_\nu + i(A^1{}_\nu + A^2{}_\nu + A_E{}^1{}_\nu + A_E{}^2{}_\nu + W^1{}_\nu + W^2{}_\nu + U^1{}_\nu + U^2{}_\nu + V^1{}_\nu + V^2{}_\nu)]F^1{}_{\sigma\mu}D^\nu R^{2\sigma\mu}\} + \\
+ a\{&-F_E{}^1{}_{\sigma\mu}(F_W{}^{2\sigma\mu} + F_U{}^{2\sigma\mu} + F_V{}^{2\sigma\mu} + F^{2\sigma\mu}) - F^1{}_{\sigma\mu}(F_E{}^{2\sigma\mu} + F_W{}^{2\sigma\mu} + F_U{}^{2\sigma\mu} + F_V{}^{2\sigma\mu}) -
\end{aligned}$$

---

[131] We show gravity weakens at very short distances using our Two-Tier Quantum Field Theory formalism. See Blaha (2003) and (2005a) amog other books by the author.

[132] We only consider the gauge field lagrangian terms in this chapter and in this book.

[133] The coupling constants of the gauge fields are not shown in the interests of simplicity. See eq. 4.83 for the coupling constants, which will be treated as implicit in the gauge fields in the remainder of this chapter.

$$- F_W{}^1{}_{\sigma\mu}(F_E{}^{2\sigma\mu} + F_U{}^{2\sigma\mu} + F_V{}^{2\sigma\mu} + F^{2\sigma\mu}) + (F_U{}^1{}_{\sigma\mu} + F_V{}^1{}_{\sigma\mu})(F_E{}^{2\sigma\mu} + F^{2\sigma\mu}) +$$
$$+ iR^1{}_{\sigma\mu}(F_E{}^{2\sigma\mu} + F_W{}^{2\sigma\mu} + F_U{}^{2\sigma\mu} + F_V{}^{2\sigma\mu} + F^{2\sigma\mu}) +$$
$$+ i(F_E{}^1{}_{\sigma\mu} + F_W{}^1{}_{\sigma\mu} + F_U{}^1{}_{\sigma\mu} + F_V{}^1{}_{\sigma\mu} + F^1{}_{\sigma\mu})R^{2\sigma\mu}\}] \qquad (12.32)$$

Thus $\mathcal{L}_{SU(3)}$, $\mathcal{L}_{SU(2)}$, $\mathcal{L}_U$, $\mathcal{L}_V$, $\mathcal{L}_E$ and $\mathcal{L}_{int}$ are the dominant interactions within hadrons, and $\mathcal{L}_G$, $\mathcal{L}_E$ and $\mathcal{L}_{int}$ are the dominant interactions in space within the framework of this discussion.

The $D_\nu R^1{}_{\sigma\mu}$ and $D^\nu R^{2\sigma\mu}$ terms have the form:

$$D_\nu R^i{}_{\sigma\mu} = + \partial_\nu R^i{}_{\sigma\mu} \; - H^{1\beta}{}_{\sigma\nu} R^i{}_{\beta\mu} - H^{1\beta}{}_{\nu\mu} R^i{}_{\sigma\beta}$$

for i = 1, 2.

The basis of the unification in the seven interaction Riemann-Christoffel curvature tensor leads to interactions beyond those in The Standard Model. These additional interactions in $\mathcal{L}_{int}$ lead to new phenomena in this unified theory such as: 1) a possible relationship between the coupling constants appearing below; 2) a possible relationship between parts of these interactions, 3) a possible solution of the proton spin puzzle due to electromagnetic-gluon terms in $\mathcal{L}_{int}$ that have not been previously considered, 4) a possible explanation of the proton radius puzzle, and other results such as explaining the different experimental results for the proton radius.

We now introduce a strong interaction coupling constant f, an electromagnetic coupling constant e, Weak coupling constant g_W, Generation group coupling constant g_G, and Layer group coupling constant g_V with[134]

$$
\begin{aligned}
A^1{}_\mu &\rightarrow fA^{1\mu} \qquad (12.33)\\
A^2{}_\mu &\rightarrow fA^{2\mu}\\
A_E{}^1{}_\mu &\rightarrow eA_E{}^{1\mu}\\
A_E{}^2{}_\mu &\rightarrow eA_E{}^{2\mu}\\
W^1{}_\mu &\rightarrow g_W W^{1\mu}\\
W^2{}_\mu &\rightarrow g_W W^{2\mu}\\
U^1{}_\mu &\rightarrow g_G W^{1\mu}\\
U^2{}_\mu &\rightarrow g_G W^{2\mu}\\
V^1{}_\mu &\rightarrow g_V W^{1\mu}\\
V^2{}_\mu &\rightarrow g_V W^{2\mu}
\end{aligned}
$$

leading to

$$
\begin{aligned}
F^1{}_{\kappa\mu} &= \partial A^1{}_\mu/\partial x^\kappa - \partial A^1{}_\kappa/\partial x^\mu + if[A^1{}_\kappa, A^1{}_\mu] \qquad (12.34)\\
F^2{}_{\kappa\mu} &= \partial A^2{}_\mu/\partial x^\kappa - \partial A^2{}_\kappa/\partial x^\mu + if[A^2{}_\kappa, A^2{}_\mu] + if[A^1{}_\kappa, A^2{}_\mu] + if[A^2{}_\kappa, A^1{}_\mu]\\
F_W{}^1{}_{\kappa\mu} &= \partial W^1{}_\mu/\partial x^\kappa - \partial W^1{}_\kappa/\partial x^\mu + ig_W[W^1{}_\kappa, W^1{}_\mu]\\
F_W{}^2{}_{\kappa\mu} &= \partial W^2{}_\mu/\partial x^\kappa - \partial W^2{}_\kappa/\partial x^\mu + ig_W[W^2{}_\kappa, W^2{}_\mu] + ig_W[W^1{}_\kappa, W^2{}_\mu] + ig_W[W^2{}_\kappa, W^1{}_\mu]\\
F_U{}^1{}_{\kappa\mu} &= \partial U^1{}_\mu/\partial x^\kappa - \partial U^1{}_\kappa/\partial x^\mu + ig_G[U^1{}_\kappa, U^1{}_\mu]\\
F_U{}^2{}_{\kappa\mu} &= \partial U^2{}_\mu/\partial x^\kappa - \partial U^2{}_\kappa/\partial x^\mu + ig_G[U^2{}_\kappa, U^2{}_\mu] + ig_G[U^1{}_\kappa, U^2{}_\mu] + ig_G[U^2{}_\kappa, U^1{}_\mu]
\end{aligned}
$$

---

[134] The coupling constants are assumed to be the renormalized physical coupling constants.

$$F_{V\,\kappa\mu}^{1} = \partial V^1_{\mu}/\partial x^{\kappa} - \partial V^1_{\kappa}/\partial x^{\mu} + ig_V[V^1_{\kappa}, V^1_{\mu}]$$
$$F_{V\,\kappa\mu}^{2} = \partial V^2_{\mu}/\partial x^{\kappa} - \partial V^2_{\kappa}/\partial x^{\mu} + ig_V[V^2_{\kappa},V^2_{\mu}] + ig_V[V^1_{\kappa},V^2_{\mu}] + ig_V[V^2_{\kappa},V^1_{\mu}]$$

Then the Strong Interaction lagrangian density terms are modified to:[135,136]

$$\mathcal{L}_{SU(3)} = \text{Tr}\,\sqrt{g}[MD_{SU(3)\nu}R'^{1}_{SU(3)\sigma\mu}D_{SU(3)}{}^{\nu}R'^{2}_{SU(3)}{}^{\sigma\mu} + aR'^{1}_{SU(3)\sigma\mu}R'^{2}_{SU(3)}{}^{\sigma\mu} - dA^2_{\mu}A^{2\mu}] \quad (12.35)$$

with

$$D_{SU(3)\nu} = [\partial_\nu + if(A^1_{\nu} + A^2_{\nu})]$$

and the electromagnetic lagrangian density term is now

$$\mathcal{L}_E = \sqrt{g}\{aF^1_{E\sigma\mu}F^{2}_{E}{}^{\sigma\mu}\} \quad (12.36)$$

Corresponding changes take place in $\mathcal{L}_{SU(2)}$ and $\mathcal{L}_{int}$.

We now approximate the metric determinant as g = 1 within hadrons. Thus the Strong lagrangian part becomes

$$\mathcal{L}_{SU(3)} = \text{Tr}\,\{MD_\nu R'^{1}_{SU(3)\sigma\mu}D^{\nu}R'^{2}_{SU(3)}{}^{\sigma\mu} + aR'^{1}_{SU(3)\sigma\mu}R'^{2}_{SU(3)}{}^{\sigma\mu} - dA^2_{\mu}A^{2\mu}\} \quad (12.37)$$
$$= \text{Tr}\,\{MD_\nu R'^{1}_{SU(3)\sigma\mu}D^{\nu}R'^{2}_{SU(3)}{}^{\sigma\mu} + \zeta R'^{1}_{SU(3)\sigma\mu}R'^{2}_{SU(3)}{}^{\sigma\mu} - \varsigma A^2_{\mu}A^{2\mu}\} \quad (12.38)$$

where

$$D_\nu R'^{j}_{SU(3)\sigma\mu} = \partial_\nu R'^{j}_{SU(3)\sigma\mu} + f[A^1_{\nu}, R'^{j}_{SU(3)\sigma\mu}] \quad (12.39)$$

for j = 1, 2.

12.5 The Vector Boson and Graviton Interaction Terms

In principle there are 42 two interaction 'interactions' amongst the seven interactions. There are 210 three interaction 'interactions' among triplets of interactions. However the actual number of interactions between interactions is significantly reduced by taking traces. In addition the interactions of the Generation Group vector bosons, and the Layer group vector bosons, with the Strong interaction and electromagnetism appear to be their only significant interactions due to the smallness of coupling constants of the other interactions.

The interaction terms of the six[137] interactions in the unified lagrangian are given by eq. 12.32 modulo the interaction with gravitation due to the overall metric determinant factor $\sqrt{g}$. In this chapter we will describe the interactions involving the Generation group and the Layer group. The interactions amongst the four known interactions are presented in chapter 9.

---

[135] The form is virtually identical to S. Blaha, Phys. Rev. D**11**, 2921 (1974), and Blaha's 1976 Gravity Research Foundation Essay, except for the initial derivative term. See Appendices A and D.

[136] We note the constant a, which appears in this chapter and chapter 5 is NOT the Charmonium constant a in eq. 2.1.

[137] The seventh interaction – the General Coordinate U(4) Reality group interaction – is discussed in chapter 13.

$\mathcal{L}_{int}$ has electromagnetic-Strong interaction terms, gravitation-Strong interaction terms, and gravitation-electromagnetic terms, Weak-Strong terms, Weak-electromagnetic terms, and Weak-Gravitation terms plus interactions with the General Coordinate U(4) Reality group fields, the Generation group fields, and the Layer group fields. Restricting consideration to the interaction of the Generation group fields, and the Layer group fields to interactions with the electromagnetic field and the Strong interaction fields, since the other possible interactions are not expected to be significant due to their small coupling constants, we approximate $\mathcal{L}_{int}$ with

$$\mathcal{L}_{int} \cong \mathcal{L}_{intE\text{-}S} + \mathcal{L}_{intE\text{-}G} + \mathcal{L}_{intS\text{-}G} + \mathcal{L}_{intS\text{-}W} + \mathcal{L}_{intG\text{-}W} + \mathcal{L}_{intE\text{-}W} + \mathcal{L}_{intS\text{-}U} + \mathcal{L}_{intE\text{-}U} + \mathcal{L}_{intS\text{-}V} + \mathcal{L}_{intE\text{-}V} \tag{12.40}$$

Since $\mathcal{L}_{int}$ is a trace it simplifies significantly. With these points in mind we proceed to consider interactions between vector bosons of differing interactions. The interactions involving the Generation group interaction and Layer group interaction in $\mathcal{L}_{int}$ neglecting self-interactions due to quadratic and triplet terms in the field terms are:[138,139]

12.5.1 Generation Group Interaction–Strong Interaction Interaction Terms

The terms in eq. 12.32 that generate the interplay of these interactions are

$$\mathcal{L}_{intS\text{-}G} = 2\ \mathrm{Tr}\ \sqrt{g}M({U^1}_\nu + {U^2}_\nu){F^1}_{\sigma\mu}(U^{1\nu} + U^{2\nu})F^{2\sigma\mu} \tag{12.41}$$

These terms do not appear in The Standard Model as it is commonly depicted.

12.5.2 Generation Group Interaction–Electromagnetic Interaction Effects

The terms in eq. 12.32 that generate the interplay of these interactions are

$$\mathcal{L}_{intE\text{-}G} = \mathrm{Tr}\ \sqrt{g}a\{-{F_E}^1{}_{\sigma\mu}{F_U}^{2\sigma\mu} + {F_U}^1{}_{\sigma\mu}{F_E}^{2\sigma\mu}\} = 0 \tag{12.42}$$

These interaction terms are zero upon taking the trace.

12.5.3 Layer Group Interaction–Strong Interaction Effects

The terms in eq. 12.32 that generate the interplay of these interactions are

$$\mathcal{L}_{intS\text{-}L} = 2\ \mathrm{Tr}\ \sqrt{g}M({V^1}_\nu + {V^2}_\nu){F^1}_{\sigma\mu}(V^{1\nu} + V^{2\nu})F^{2\sigma\mu} \tag{12.43}$$

These terms do not appear in The Standard Model as commonly depicted.

---

[138] The coupling constants of the gauge fields are not shown in the interests of simplicity. See eq. 12.33 for the coupling constants, which will be treated as implicit in the gauge fields in this chapter.

[139] It is understood that the traces are taken separately for SU(2), SU(3) and U(4) matrices.

12.5.4 Layer Group Interaction–Electromagnetic Interaction Effects

The terms in eq. 12.32 that generate the interplay of these interactions are zero after taking the trace.

$$\mathcal{L}_{\text{intE-L}} = 0$$

13. The Seventh Interaction – Gauge Field Form of General Coordinate U(4) Reality Group Transformations

The seventh interaction emerges from the transition from complex General Relativistic transformations. In section 11.2.10 we separated the set of all complex General Coordinate transformations into 1) by factoring each transformation into two factors, a real-valued transformation and a complex-valued transformation, each set of factors form a subset, and then 2) separating the set of complex factors into those that satisfy

$$\Lambda(\omega, \mathbf{u})^T G \Lambda(\omega, \mathbf{u}) = G \tag{11.50}$$

and those that do not. We then showed that the set of those that do not satisfy eq. 11.50 form a curved space representation of the U(4) group under 'multiplication' of transformations.

The elements of the set of real and complex General Coordinate transformations whose flat complex space-time limit satisfies eq. 11.50 form the elements of the Complex Lorentz group.

We thus find the set of all 4-dimensional curved space General coordinate transformations can be visualized as in Fig. 13.1.

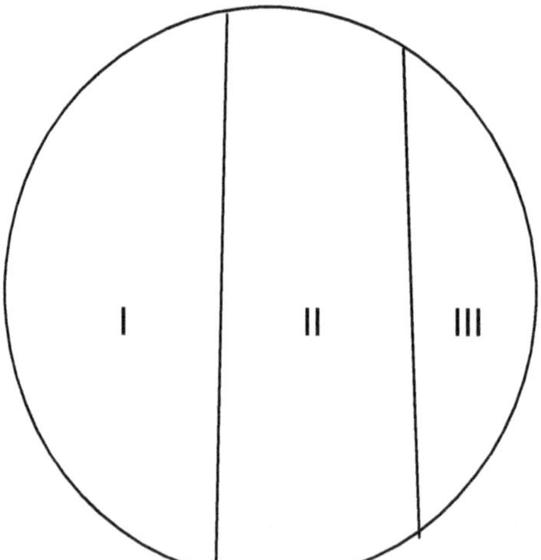

Figure 13.1. A visualization of the set of General Coordinate transformations separated into real-valued General coordinate transformations (part I), complex transformations that satisfy eq. 11.50 (part II), and complex transformations that

do not satisfy eq. 11.50 (part III). Part I and part II combine in the limit of flat space-time to form the Complex Lorentz group. Part III elements do not satisfy eq. 11.50 and form a U(4) group that we call the General Relativistic Reality group.

A complex transformation of type III has the form:

$$U(x'')^{\mu}{}_{\nu} = w^{\mu}{}_{a}(x'')\left[\exp(i\textstyle\sum_k \Phi_k(x'')\tau_k)\right]^{a}{}_{b}\, l^{b}{}_{\nu}(x'') \qquad (11.20)$$

$$U^{-1}(x'')^{\mu}{}_{\nu} = w^{\mu}{}_{a}(x'')\left[\exp(-i\textstyle\sum_k \Phi_k(x'')\tau_k)\right]^{a}{}_{b}\, l^{b}{}_{\nu}(x'') \qquad (11.21)$$

An infinitesimal transformation is approximately

$$U(x'')^{\nu}{}_{\beta} \approx \delta^{\nu}{}_{\beta} + i\textstyle\sum_k \Phi_k(x'')[\tau_k]^{\nu}{}_{\beta} \qquad (11.34)$$

$$U^{-1}(x'')^{\nu}{}_{\beta} \approx \delta^{\nu}{}_{\beta} - i\textstyle\sum_k \Phi_k(x'')[\tau_k]^{\nu}{}_{\beta} \qquad (11.35)$$

using the *vierbein* flat space-time limits

$$w^{\mu}{}_{a}(x'') \approx \delta^{\mu}{}_{a}$$

$$l^{b}{}_{\nu}(x'') \approx \delta^{b}{}_{\nu}$$

where

$$\Phi_k(x) = \int^{x} dy_{\lambda}\, A_{Rk}{}^{\lambda}(y) \qquad (11.40)$$

As a result

$$\Gamma_R{}^{\sigma}{}_{\lambda\mu} = -\tfrac{1}{2}i\{\textstyle\sum_k A_{Rk}(x'')_{\mu}[\tau_k]^{\sigma}{}_{\lambda} + \textstyle\sum_k A_{Rk}(x'')_{\lambda}[\tau_k]^{\sigma}{}_{\mu}\} \qquad (11.41)$$

$$= A_R{}^{\sigma}{}_{\mu\lambda} + A_R{}^{\sigma}{}_{\lambda\mu}$$

(summed over k) with the matrix $A_R{}^{\sigma}{}_{\mu\lambda}$ given by

$$A_R{}^{\sigma}{}_{\mu\lambda} = -\tfrac{1}{2}i\textstyle\sum_k A_{Rk\mu}[\tau_k]^{\sigma}{}_{\lambda} \qquad (11.42)$$

We note

$$R^{1\beta}{}_{\sigma\nu\mu} = \partial_{\mu}H^{1\beta}{}_{\sigma\nu} - \partial_{\nu}H^{1\beta}{}_{\sigma\mu} + H^{1\gamma}{}_{\nu\sigma}H^{1\beta}{}_{\gamma\mu} - H^{1\gamma}{}_{\mu\sigma}H^{1\beta}{}_{\gamma\nu} \qquad (12.12)$$

$$R^{2\beta}{}_{\sigma\nu\mu\rho} = \partial_{\mu}H^{2\beta}{}_{\sigma\nu} - \partial_{\nu}H^{2\beta}{}_{\sigma\mu} + H^{2\gamma}{}_{\nu\sigma}H^{2\beta}{}_{\gamma\mu} - H^{2\gamma}{}_{\mu\sigma}H^{2\beta}{}_{\gamma\nu} +$$

$$+ H^{1\gamma}{}_{\nu\sigma}H^{2\beta}{}_{\gamma\mu} - H^{1\gamma}{}_{\mu\sigma}H^{2\beta}{}_{\gamma\nu} + H^{2\gamma}{}_{\nu\sigma}H^{1\beta}{}_{\gamma\mu} - H^{2\gamma}{}_{\mu\sigma}H^{1\beta}{}_{\gamma\nu} \qquad (12.13)$$

where

$$H^{\sigma}{}_{\nu\mu} = \Gamma_{GR}{}^{\sigma}{}_{\nu\mu} + \Gamma_{GR}{}^{2\sigma}{}_{\nu\mu} + \Gamma_R{}^{1\sigma}{}_{\nu\mu} + \Gamma_R{}^{2\sigma}{}_{\nu\mu} \qquad (11.47)$$

In flat space-time

$$\Gamma_R{}^{\sigma}{}_{\lambda\mu} = A_R{}^{\sigma}{}_{\mu\lambda} + A_R{}^{\sigma}{}_{\lambda\mu} \qquad (13.1)$$

with $A_R{}^{\sigma}{}_{\mu\lambda}$ transformable to matrix row and column numbers

$$A_{Rflat}{}^{i\mu a}{}_{b} = A_{Rflat}{}^{i\mu}[\tau_k]^{\sigma}{}_{\lambda}\delta_{\sigma}{}^{a}\delta^{\lambda}{}_{b} \qquad (11.48)$$

And so $A_{Rflat}{}^{ia}{}_{\mu b}$ may be written in matrix form as

$$A_{Rflat}{}^{i}{}_{\mu} = -\tfrac{1}{2}i\sum_k A_{Rflat}{}^{i}{}_{k\mu}\tau_k \qquad (11.49)$$

The relevant *quadratic* $A_R{}^{\sigma}{}_{\mu\lambda}$ terms from eq. 12.29 that are needed to find the dynamic equation for the gauge fields $A_{Rflat}{}^{i}{}_{\mu}$ are contained in

$$\mathcal{L}_A = \text{Tr}\,\sqrt{g}[M\partial_\nu R^1{}_{\sigma\mu}\partial^\nu R^{2\sigma\mu} + aR^1{}_{\sigma\mu}R^{2\sigma\mu} + bg^{\sigma\mu}(R^1{}_{\sigma\mu} + R^2{}_{\sigma\mu}) + 1/4(g_{\mu\nu} + g^2{}_{\mu\nu})T^{\mu\nu}] \qquad (13.2)$$

with c = c' = 0 as in chapter 4. We can

$$R^i{}_{\sigma\mu} = R^{i\beta}{}_{\sigma\beta\mu} \equiv \partial_\mu(A_R{}^{i\beta}{}_{\sigma\beta} + A_R{}^{i\beta}{}_{\beta\sigma}) - \partial_\beta(A_R{}^{i\beta}{}_{\sigma\mu} + A_R{}^{i\beta}{}_{\mu\sigma}) \qquad (13.3)$$

for i = 1, 2. In the flat space-time limit we chose the Landau gauge

$$\partial_\mu A_{Rflat}{}^{i\mu a}{}_{b} = 0 \qquad (13.4)$$

As a result

$$R^i{}_{\sigma\mu} \equiv \partial_\mu(A_R{}^{i\beta}{}_{\sigma\beta} + A_R{}^{i\beta}{}_{\beta\sigma}) \qquad (13.5)$$

Using eq. 11.48 and

$$\begin{aligned} A_{Rflat}{}^{i\mu\sigma}{}_{\lambda} &= A_{Rflat}{}^{i\mu}[\tau_k]^{a}{}_{b}\delta^{\sigma}{}_{a}\delta_{\lambda}{}^{b} \\ A_{Rflat}{}^{i}{}_{\mu} &= -\tfrac{1}{2}i\sum_k A_{Rflat}{}^{i}{}_{k\mu} \end{aligned} \qquad (13.6)$$

and taking the trace in eq. 13.2 we obtain

$$\mathcal{L}_A = \text{Tr}\,\sqrt{g}[8M\partial_\nu\partial_\mu A_{Rflat}{}^{1}{}_{\sigma}\partial^\nu\partial^\mu A_{Rflat}{}^{2\sigma} + 8a\partial_\mu A_{Rflat}{}^{1}{}_{\sigma}\partial^\mu A_{Rflat}{}^{2\sigma} + 1/4(g_{\mu\nu} + g^2{}_{\mu\nu})T^{\mu\nu}] \qquad (13.7)$$

in the flat space-time limit. Eq. 13.2 needs to take account of the complex nature of $(g_{\mu\nu} + g^2{}_{\mu\nu})$ until transformed by the infinitesimal form of the complex Reality transformation of eqs. 11.20 and 11.21 above.

$$\begin{aligned} (g_{\beta\alpha} + g^2{}_{\beta\alpha})' &\to U(x'')_\beta{}^\mu(g_{\mu\nu} + g^2{}_{\mu\nu})U^{-1}(x'')^\nu{}_\alpha \\ &= (\delta_\beta{}^\mu + i\sum_k\Phi_k(x'')[\tau_k]_\beta{}^\mu)(g_{\mu\nu} + g^2{}_{\mu\nu})(\delta^\nu{}_\alpha - i\sum_k\Phi_k(x'')[\tau_k]^\nu{}_\alpha) \end{aligned}$$

$$\cong (g_{\beta\alpha} + g^2{}_{\beta\alpha}) + i\{\sum_k \Phi_k(x'')[\tau_k]_\beta{}^\mu - i \sum_k \Phi_k(x'')[\tau_k]^\nu{}_\alpha)(g_{\mu\nu} + g^2{}_{\mu\nu})$$

$$\cong (g_{\beta\alpha} + g^2{}_{\beta\alpha}) + i\sum_k \Phi_k(x'')\{[\tau_k]_{\beta\alpha} - [\tau_k]_{\beta\alpha}\} \quad (13.8)$$

Approximating $\Phi_k(x'')$ with an infinitesimal line we find

$$\sum_k \Phi_k(x) \cong \delta x_\lambda A_{Rflat}{}^{1\lambda}(x) \quad (13.9)$$

by eq. 11.40. Thus

$$1/4(g_{\mu\nu} + g^2{}_{\mu\nu})T^{\mu\nu} \cong ¼ [(g_{\beta\alpha} + g^2{}_{\beta\alpha}) + i\, \delta x_\lambda A_{Rflat}{}^{1\lambda}(x)\{[\tau_k]_{\beta\alpha} - [\tau_k]_{\beta\alpha}\}]T^{\alpha\beta}$$

Applying the canonical Euler-Lagrange method we obtain the dynamical equations (using integration by parts as in chapter 4 to handle higher order derivative terms):[140]

$$\Box^2 A_{Rflat}{}^1{}_\sigma + (a/M)\Box A_{Rflat}{}^1{}_\sigma + i\delta x_\sigma\{[\tau_k]_{\beta\alpha} - [\tau_k]_{\beta\alpha}\}T^{\alpha\beta}/(32M) = 0 \quad (13.10)$$

$$\Box^2 A_{Rflat}{}^2{}_\sigma + (a/M)\Box A_{Rflat}{}^2{}_\sigma = 0 \quad (13.11)$$

Since the A_{Rflat} gauge field is gravitational in nature it exists, as eq. 13.10 shows, as a type of gravitational interaction whose source is the energy-momentum tensor. Following the derivation of the gravitational potential in section 6.2 we find the Coulomb interaction of $A_{Rflat}{}^{10}$.

13.1 Gravity Potential

Assuming that we are dealing with non-relativistic matter we can calculate the gravity potential contribution from eq. 13.10:

$$V_{GA1}(\mathbf{x}) = -(1/32M)\int d^3k \exp(i\mathbf{k}\cdot\mathbf{x})V_{GA1}(\mathbf{k})/(2\pi)^3 \quad (13.12)$$

where

$$V_{GA1}(\mathbf{k}) = (\mathbf{k}^4 + (a/M)\mathbf{k}^2)^{-1} \quad (13.13)$$

The eq. 13.13 can be separated into two terms:

$$V_{GA1}(\mathbf{k}) = (M/a)[1/k^2 - 1/(\mathbf{k}^2 + a/M)] \quad (13.14)$$

which yield

$$V_{GA1}(\mathbf{r}) = -[1/(96\pi a)][1/r - e^{-m_A r}/r] \quad (13.15)$$

---

[140] It is possible that the Reality transformation also depends on $A_{Rflat}{}^2{}_\sigma$. Then eq. 13.11 would have an energy-momentum tensor term as well. Consequently there would be an additional interaction of the same form as eq. 13.12.

In chapter 10 we found

$$M \sim 2\times10^{161}\ \text{ev}^{-2} \tag{10.12}$$

with the result

$$m_A = (a/M)^{½} = m_{SI} \cong 10^{-71}\ \text{GeV} \tag{13.16}$$

by eq. 5.20. Since $a \cong 1$ the coupling constant

$$1/(96\pi a) \cong 0.0033 \tag{13.17}$$

In comparison the electromagnetic fine structure constant is

$$\alpha \cong 0.0073$$

Thus the A_R coupling constant is approximately ½ of the fine structure constant. However expanding eq. 13.15 in r for 'small' distances (much less than galactic distances) we find

$$V_{GA1}(\mathbf{r}) = -[1/(96\pi a)](m_A + m_A^2 r) \tag{13.18}$$

yielding an attractive force

$$\mathbf{F}_A \cong -[m_A^2/(96\pi a)]\mathbf{r}/r = -G_A\ \mathbf{r}/r \tag{13.19}$$

where

$$G_A = [m_A^2/(96\pi a)] \approx 10^{-142}\ \text{GeV}^2$$

in comparison to the Planck mass *squared* of the order of 10^{38} GeV^2. Thus the Gravitation gauge field force appears to be extremely small in comparison to the other forces of Nature.

13.2 Influence of Gravitational Gauge Field on Gravitation

The gravitational gauge field potential in eq. 13.15 has a relatively large coupling constant that makes it competitive with the known force of gravity at large distances of the scale of galactic distances. This force, which is negligible at short distances of the order of planetary distances, may be part of the MoND phenomena that affects the motion of stars.

At large galactic scale distances the gravitational gauge field potential between matter (energy-momentum density) is

$$V_{GA1}(\mathbf{r}) \cong -[1/(96\pi a)]/r \tag{13.20}$$

with $1/(96\pi a) \cong 0.0033$ giving a formidable contribution to total gravity. At short distances of the order of solar system scale distances

$$V_{GA1}(\mathbf{r}) \cong 0 \tag{13.21}$$

which may account for its non-detection in earth-based gravity experiments.

13.3 Does Unusual Gravitational Force at Galactic Distances Imply Coordinates are Complex?

If the existence of the gravitational gauge field could be determined experimentally then it would seem to be strong evidence that our universe has complex coordinates that are mapped to real-valued coordinates by the Reality groups.

The reason that we do not know of the complex nature of coordinates is simply the absence of complex valued measuring sticks. Rulers always yield real values for distances. Clocks only measure real-valued times.

Complex space-time coordinates opens the doors to a variety of new phenomena and to new aspects of known phenomena such as Black Holes.

A complex space-time would imply complex Lorentz transformations between coordinate systems – the Complex Lorentz group[141] *– and faster-than-light particle velocities. The author has advocated the existence of faster-then-light velocities for many years. There is supporting experimental evidence for this possibility, which is cited in earlier books by the author.*

---

[141] As used in Streater and Wightman (2000).

14. Higher Derivative Theories and Boson Interaction Unification

14.1 MoND or a Higher Derivative Theory of Elementary Particles

In this book we have achieved a meaningful unification of the Strong Interactions and Gravitation that unites major properties of both: quark confinement and MoND (Modified Newtonian Gravity) without sacrificing the fundamental principles of dynamics: Newtonian dynamics, quantum theory and Quantum Field Theory. In contrast the MoND proposal requires a major modification of all these classical and quantum theories – indeed the complete panoply of Modern Physics.

Given a choice between the two approaches it is clear that our approach must be viewed as the correct one. Perhaps the best proof of this statement is the incredible accuracy of QED calculations by Kinoshita, his predecessors, and his colleagues. It is inconceivable that such accuracy could be obtained by a very different formulation of Physics such as the original and recent MoND proposals.

14.2 Impact of Unification based on the Riemann-Christoffel Curvature Tensor

The unification of the seven elementary particle interactions and gravitation achieved in this book was based on the construction of the Riemann-Christoffel curvature tensor for all these interactions. The physical motivation for this approach is the simple fact that all interactions involve energy and momentum and, in principle, they all influence the curvature of space to a greater or lesser degree.

The unification of boson gauge fields based on the Riemann-Christoffel symbol complements the 'unification' of the spectrum of fermions based on the Complex Lorentz group, and the U(4) Generation group and the U(4) Layer group, which is described in Blaha (2015a), (2016a), (2016b) and (2016c) as well as in the author's earlier books.

Appendix A. The Strong Interaction and Charmonium

This appendix appeared as chapter 2 of Blaha (2016g). It is reproduced here because its form directly relates to the Strong Interaction subsector of the new unified theory presented here.

As a preliminary to the development of our unified GEMS theory we will consider the Strong Interaction potential used in studies of Charmonium bound states.[142] This potential will be seen to be that of our Strong Interaction theory developed some years ago with one simple modification. Subsequently we will develop the structure of the unified theory in some detail.

In 1974 a bound state of a charmed and an anti-charmed quark was discovered by two experimental groups. Since charmed quarks are quite massive, theoretical attempts were made to understand the charmed quark bound states within the framework of non-relativistic quantum mechanics. The "Cornell group" developed a fairly satisfactory[143] charmed quark bound state spectrum in 1974-5 using a combination of a linear and a 1/r potential as the strong interaction. In a recent fit[144] they gave the potential energy:

$$V(r) = -\kappa/r + r/a^2 \qquad \text{(A-2.1)}$$

where $\kappa = 0.61$, $a = 2.38\ \text{GeV}^{-1}$ and the charmed quark mass was 1.84 GeV.

A-2.1 The Origin of the Linear Potential

The linear potential appears to have originated in a suggestion of Feynman in the Spring of 1974. This author proposed[145] a non-Abelian gauge quantum field theory, which yielded a linear potential. These papers, which had 4th order dynamic equations for the gauge fields, showed how to avoid the problems previously associated with higher derivative theories by using principal-value gauge field propagators that were similar in concept to the action-at-a-distance propagators used by Feynman and Wheeler in the late 1940's to formulate action-at-a-distance Electrodynamics.

---

[142] This chapter first appeared in Blaha (2016a).
[143] As did a Harvard group.
[144] E. J. Eichten, K. Lane, and C. Quigg, arXiv:hep-ph/ 0206018 (2002). See this paper for references to earlier work by the "Cornell group" and the "Harvard group" as well as papers by other researchers.
[145] S. Blaha, Phys. Rev. D**10**, 4268 (July, 1974) and Phys. Rev. D**11**, 2921 (December, 1974). They appear in Appendices A and B of this book for the reader's convenience. These papers appeared before the charmonium calculations of the Cornell and Harvard groups in 1975.

Thus we created a non-Abelian quantum field theory of the strong interaction yielding a linear potential. In parallel with this development, Professor Kenneth Wilson (later a Nobelist) was developing lattice gauge theory. Because lattice lines focus the field of gauge bosons, lattice gauge theory also exhibited a linear potential between quarks. Thus it offered an alternative to our gauge theory. However, the linearity of the lattice potential was "built-in" by the lattice theory formulation and thus was an artifact of the lattice formulation. This approach and other proposed approaches all share the problem that the linear potential that they produce cannot be proven to truly be a consequence. Rather the linear potential is the "likely result."

On the other hand our higher dimensional theory produces the linear potential if the standard rules of quantum field theory are followed with the proviso that gauge field propagators are principal-value propagators.

This author had many discussions with Professor Wilson in late 1974 in the author's office and while walking to lunch at the Cornell Faculty Club. Professor Wilson proposed possible flaws in the author's theory on an almost daily basis. The author was able to show these possible flaws were not flaws but physically acceptable. The final discussion with Wilson ended with Wilson stating words to the effect, "Your theory may be a correct phenomenological approximation to my theory of the strong interaction and quark confinement. But my theory is the correct one. Your theory is only a phenomenology." In the forty plus years since this concluding discussion no one has proved that the conventional strong interaction theory truly has a linear potential and quark confinement although some approximations suggest it does.

In the absence of a demonstration of a linear potential in the standard strong interaction model we suggest our theory is a viable alternative. Since the linear potential appears to fairly successfully describe much of the charmonium spectrum we feel our theory with its explicit derivation of a linear potential is worthy of interest – especially because it is in agreement with experiment as far as we know. *An experimentally completely correct phenomenology is a theory.*

A-2.2 Charmonium Potential in the Light of our Theory

The author's non-abelian strong interaction theory presented in Appendices A and B needs one modification to yield the charmonium potential eq. A-2.1. In Appendix A eqs. 6 and 18 should have the interaction term expanded:

$$g\mathcal{A}_1 \rightarrow g(\mathcal{A}_1 + \mathcal{A}_2) \tag{A-2.2}$$

and similarly in eq. 20 in Appendix B.

Eqs. 38 – 41 of Appendix A directly show that the additional interaction term leads to a gluon propagator[146] $<A^1 + A^2, A^1 + A^2> = 2<A^1, A^2> + <A^1, A^1>$, and introduces a 1/r term in the potential part of the propagator. This leads to a potential of the form of eq. A-2.1 with $g = \sqrt{(\kappa/2)} = 0.552$ and $\lambda = 0.761$ GeV. Thus

[146] Eqs. 40-41 in Appendix A.

$$V(r) = -2g^2/r + g^2\lambda^2 r \qquad \text{(A-2.3)}$$

The charmonium potential emerges directly from our theory. We note that $g^2/4\pi = 0.024$ is only a factor of 3.3 more than the fine structure constant – approximately 1/137 = 0.0073. *Therefore perturbative corrections to our inter-quark potential may not be significant and our theory may be the correct theory of the strong interaction. The strong interaction potential in this charmonium fit suggests that the unperturbed potential of the theory presented in Appendices A and B may be a good approximation to the exact potential determined in perturbation theory.*

The smallness of the Strong Interaction terms in the Cornell group potential (eq. A-2.1) looks puzzling at first glance. Why is it not large ("Strong")? We believe the strength of the Strong Interaction does not originate in the value of the coupling constant but rather in the linear potential term which provides confinement. The Cornell group potential's 'small' coupling constant then becomes understandable within the context of Strong Interaction phenomenology.

A-2.3 Asymptotic Freedom in Our Model of the Strong Interaction

The combined gluon propagator in momentum space in our theory has the form

$$-1/k^2 + \lambda^2/k^4 \qquad \text{(A-2.4)}$$

In the high energy (short distance) limit k → ∞ it approaches k^{-2} yielding the same short distance behavior as the usual strong interaction model. In the low energy (long distance) limit k → 0 it has a confining k^{-4} behavior. Thus we explicitly obtain quark confinement and asymptotic freedom in our model.

Appendix B. Order of Magnitude Commonality of the Coupling Constants of the Four Known Interactions

This appendix contains an extract from Blaha (2015c) that shows that if interaction coupling constants are generated from a Higgs-like mechanism (as particle masses are), then they are of the same order of magnitude – order of magnitude unity. This result depends on the Planck mass being used as the common scale.

Beginning of Extract

In Blaha (2015c) we introduced a new formalism for the generation of non-zero vacuum expectation values using a form of second quantization, called pseudoQuantization, developed by the author[147] in 1978 that combines both quantum and classical fields within the same framework. In this extended quantum field theory vacuum expectation values are due to coherent vacuum ground states. This approach has the advantage of 1) resolving the issue of negative energy boson states that has existed for many years;[148] and 2) showing why inertial reference frames are of special significance – a long standing question in Theoretical Physics.[149]

1.4.1 PseudoQuantization of Higgs Particles

We will now consider the pseudoQuantization of a scalar particle field that will become a particle with a non-zero vacuum expectation value. (Section III of our paper in Appendix C contains additional detail.) We begin by defining two fields that correspond to a single scalar particle: $\varphi_1(x)$ and $\varphi_2(x)$.[150] These fields will be assumed to have the equal time commutators

$$[\varphi_i(x), \pi_j(y)] = i(1 - \delta_{ij})\delta^3(\mathbf{x} - \mathbf{y}) \tag{1.1}$$
$$[\varphi_i(x), \varphi_j(y)] = 0$$
$$[\pi_i(x), \pi_j(y)] = 0$$

---

[147] This section is based on our 1978 paper that appeared in the peer-reviewed journal *Physical Review D*. The paper is reproduced in Appendix D for the reader's convenience with the kind permission of The American Physical Society. The paper also does present a new formulation of Quantum Mechanics that incorporates both quantum and classical mechanics within one framework. Recently, experimenters have been investigating the possibility of macroscopic quantum phenomena. The new formulation is ideally suited for tracing the change from a quantum to a classical regime. It also is applicable to "large n atoms" where the outermost electrons approach classical behavior with an almost continuous energy spectrum.

[148] P. A. M. Dirac once claimed to have a solution for this problem but he did not reveal it. See Blaha (2015c).

[149] Their significance follows from the existence of a common rest frame for all Higgs particles vacuum states. All other inertial reference frames result from Lorentz transformations. See Blaha (2015c),

[150] The subscripts on the fields are not gauge symmetry indices but simply identifiers distinguishing the fields from one another.

where δ_{ij} is the Kronecker δ and where $\pi_i(x)$ is the canonically conjugate momentum to $\varphi_i(x)$. The fields $\varphi_1(x)$ and $\pi_1(y)$ will be observable classical fields as shown by eqs. 69 and 70 in Appendix C. The fields $\varphi_2(x)$ and $\pi_2(y)$ will not be observables so that $\varphi_1(x)$ and $\pi_1(y)$ can both be sharp on the set of physical states.

We now specify the lagrangian density for a scalar Klein-Gordon particle:

$$\mathcal{L} = \partial\varphi_1/\partial x_\mu \partial\varphi_2/\partial x^\mu - m^2\,\varphi_1\varphi_2 \tag{1.2a}$$

and hamiltonian density

$$\mathcal{H} = \pi_1\,\pi_2 + \partial\varphi_1/\partial x_i \partial\varphi_2/\partial x^i + m^2\,\varphi_1\varphi_2 \tag{1.2b}$$

where i labels spatial coordinates, m is the particle mass,[151] and $\pi_1 = \partial\varphi_2/\partial t$ and $\pi_2 = \partial\varphi_1/\partial t$. The fields can be fourier expanded in terms of creation and annihilation operators:

$$\varphi_i(\mathbf{x}, t) = \int d^3k\,[a_i(k)f_k(x) + a_i^\dagger(k)f_k{}^*(x)] \tag{1.3}$$

for i = 1, 2 where

$$f_k(x) = e^{-ik\cdot x}/(2\omega_k(2\pi)^3)^{1/2}$$

with ω_k being the energy.

The creation and annihilation operators satisfy the commutation relations:

$$[a_i(k), a_j^\dagger(k')] = (1 - \delta_{ij})\delta^3(\mathbf{k} - \mathbf{k'}) \tag{1.4}$$
$$[a_i(k), a_j(k')] = 0$$
$$[a_i^\dagger(k), a_j^\dagger(k')] = 0$$

for i, j = 1, 2.

In this formulation the defining properties of a coherent physical state are:

$$\varphi_1(x)|\Phi, \Pi\rangle = \Phi(x)|\Phi, \Pi\rangle \tag{1.5}$$
$$\pi_1(x)|\Phi, \Pi\rangle = \Pi(x)|\Phi, \Pi\rangle$$

where $\Phi(x)$ and $\Pi(x)$ are sharp on the states and thus classical fields expressible as

$$\Phi(\mathbf{x}, t) = \int d^3k\,[\alpha(k)f_k(x) + \alpha^*(k)f_k{}^*(x)] \tag{1.6}$$

and correspondingly for $\Pi(x)$.

1.4.2 Vacuum States for Scalar Particles with Non-Zero Vacuum Expectation Values

When we implement the mass mechanism Φ becomes a constant. We can define a set of states

$$a_1(k)|\alpha\rangle = \alpha(k)|\alpha\rangle$$
$$a_1^\dagger(k)|\alpha\rangle = \alpha^*(k)|\alpha\rangle$$

and correspondingly a set of coherent states

$$|\alpha\rangle = C\exp\{\int d^3k\,[\alpha(k)a_2^\dagger(k) + \alpha^*(k)a_2(k)]\}|0\rangle \tag{1.7}$$

where C is a normalization constant, and where the vacuum state $|0\rangle$ satisfies

$$a_1(k)|0\rangle = a_1^\dagger(k)|0\rangle = 0 \tag{1.8a}$$
$$a_2(k)|0\rangle \neq 0 \qquad\qquad a_2^\dagger(k)|0\rangle \neq 0 \tag{1.8b}$$

[151] Note the mass term here has the "correct" sign unlike the mass term in the usual Higgs potential.

The dual vacuum state satisfies

$$<0|a_2(k) = <0|a_2^\dagger(k) = 0 \quad (1.9a)$$

$$<0|a_1(k) \neq 0 \qquad\qquad <0|a_1^\dagger(k) \neq 0 \quad (1.9b)$$

Additional details on these coherent states, which differ from conventional coherent states such as those of Kibble[152] and others, can be found in Appendix C.

With this coherent state formalism, which gives purely classical fields and yet also has quantum fields through the use of φ_2 and its creation and annihilation operators, we now have the machinery to define a mass mechanism without the introduction of a potential whose origin can only be described as dubious.

For we can define a coherent state for some k as

$$|\Phi, \Pi> = C\exp\{[(2\pi)^3\omega_k/2]^{1/2}\Phi[a_2^\dagger(k) + a_2(k)]\}|0> \quad (1.10)$$

where C is a normalization constant, that yields a constant, non-zero vacuum expectation value:

$$\varphi_1(x)|\Phi, \Pi> = \Phi|\Phi, \Pi> \quad (1.11)$$

where Φ is a constant. Evaluating a fermion interaction term we find a mass term emerges[153]

$$\overline{\psi}(\varphi_1 + \varphi_2)\psi \rightarrow \overline{\psi}(\Phi + \varphi_2)\psi \quad (1.12)$$

It can generate a mass for an interaction with a gauge field of the form

$$A^\mu(\varphi_1 + \varphi_2)^2A_\mu \rightarrow A^\mu(\Phi + \varphi_2)^2A_\mu \quad (1.13)$$

It also yields a quantum field theoretic interaction that would result in the production of ElectroWeak particles from these scalar fields. (The production of Higgs particles that decay into ElectroWeak gauge particles has recently been found at CERN.)

The present formalism provides a clean way to separate the vacuum expectation value of a scalar particle from its quantum field part in contrast to the Higgs Mechanism where one has to separate a Higgs field into parts manually.

It appears that our formulation of the mass generation mechanism sheds significant light on the reason for the special prominence of inertial frames. Consider massive scalars.[154] Eqs. 1.2 describe a massive scalar particle. If the scalar is massive, then the rest frame particle "vacuum" state, eq. 1.10, which yields a non-zero expectation value, is

$$|\Phi, \Pi> = C\exp\{[(2\pi)^3m/2]^{1/2}\Phi[a_2^\dagger(\mathbf{0},m) + a_2(\mathbf{0},m)]\}|0> \quad (1.10')$$

We thus find that inertial reference frames are singled out as "special" in the sense that they are the only accessible reference frames that can be generated by a Lorentz boost/transformation from the Higgs particle rest frame. *The particles vacuum state single out the class of inertial reference frames.*

...

---

[152] T. W. B. Kibble, Jour. Math. Phys. **2**, 212 (1961).
[153] When matrix elements with a "vacuum state" such as eq. 1.10 are taken.
[154] Experiments at CERN have apparently discovered a Higgs particle with a 125 GeV/c mass.

7.1 Coupling Constants Vacuum Expectation Value Generation

The appearance of just eight fundamental coupling constants in The Theory of Everything makes them ideal candidates for replacement by eight scalar Higgs particle, vacuum expectation values. *Using vacuum expectation values leads to the remarkable conclusion that all known coupling constants, properly rewritten using our pseudoQuantum vacuum expectation value formalism, all have a value of the order of unity – even the gravitational constant* $g_{CG} = \kappa^{-1} = (4\pi G)^{-½}$.

7.1.1 Yang-Mills Coupling Constants

We will first consider the case of a generic Yang-Mills field $A^{b\mu}$ of some symmetry group, a generic fermion field ψ, and a Higgs particle with fields φ_1 and φ_2 (as defined earlier). We will replace its generic coupling constant g with Higgs fields' vacuum expectation values.[155] The initial dynamic equations are

$$\partial/\partial x_\mu F^a_{\ \mu\nu} + gf^{abc}A^{b\mu} F^c_{\ \mu\nu} = j^a_{\ \nu} \tag{7.1}$$

and

$$[i\gamma^\mu(\partial/\partial x^\mu - igA_\mu) - m]\psi(x) = 0 \tag{7.2}$$

where

$$F^a_{\ \mu\nu} = \partial/\partial x^\nu A^a_{\ \mu} - \partial/\partial x^\mu A^a_{\ \nu} + gf^{abc}A^b_{\ \mu}A^c_{\ \nu} \tag{7.3}$$

and where a, b, c are structure constant indices, g is the coupling constant, and $j^a_{\ \nu}$ is the corresponding current.

A gauge transformation has the form

$$gA'_\mu(x) = -i(\partial_\mu\Omega(x))\Omega^{-1}(x) + g\Omega(x)A_\mu(x)\Omega^{-1}(x) \tag{7.4}$$

7.1.2 C-Number Field Coupling Constant

We can replace g with fields in two ways. One way is:

$$\partial/\partial x_\mu F^a_{\ \mu\nu} + m'^{-1}\varphi_1(x)f^{abc}A^{b\mu} F^c_{\ \mu\nu} = j^a_{\ \nu} \tag{7.5}$$

and

$$[i\gamma^\mu(\partial/\partial x^\mu - im'^{-1}\varphi_1(x)A_\mu) - m]\psi(x) = 0 \tag{7.6}$$

where

$$F^a_{\ \mu\nu} = \partial/\partial x^\nu A^a_{\ \mu} - \partial/\partial x^\mu A^a_{\ \nu} + m'^{-1}\varphi_1(x)f^{abc}A^b_{\ \mu}A^c_{\ \nu} \tag{7.7}$$

with the corresponding gauge transformation rule:

$$\varphi_1(x)A'_\mu(x) = -im'(\partial_\mu\Omega(x))\Omega^{-1}(x) + \varphi_1(x)\Omega(x)A_\mu(x)\Omega^{-1}(x) \tag{7.8}$$

Using the vacuum state defined by

$$|\Phi, \Pi\rangle = C\exp\{[(2\pi)^3 m/2]^{½} m'g[a_2^\dagger(\mathbf{0},m) + a_2(\mathbf{0},m)]\}|0\rangle \tag{7.9}$$

we see the equations become eqs. 7.1-7.4 since Φ = m'g. Note m' may equal m, or may be the iota Landauer mass[156] or some other value.

---

[155] This approach is conceptually similar to that of Dicke et al for the gravitational constant G. See R. H. Dicke, Phys. Rev. **125**, 2163 (1962) and references therein. See Weinberg (1972) p. 155ff, and Misner et al (1973) p. 1070 for lucid discussions.

[156] The iota mass is a universal mass equal to the Landauer energy of a logical value. See Blaha (2015a).

Thus we have developed the first Higgs-like mechanism for purely c-number coupling constants.

7.1.3 Q-Number Field Coupling Constant

The other way to reduce coupling constants to vacuum expectation values, which we believe is preferable, is

$$\partial/\partial x_\mu F^a_{\mu\nu} + m'^{-1}(\varphi_1 + \varphi_2)f^{abc}A^{b\mu} F^c_{\mu\nu} = j^a_\nu \tag{7.10}$$

and

$$[i\gamma^\mu(\partial/\partial x^\mu - i\, m'^{-1}(\varphi_1 + \varphi_2)A_\mu) - m]\psi(x) = 0 \tag{7.11}$$

where

$$F^a_{\mu\nu} = \partial/\partial x^\nu A^a_\mu - \partial/\partial x^\mu A^a_\nu + m'^{-1}(\varphi_1 + \varphi_2)f^{abc}A^b_\mu A^c_\nu \tag{7.12}$$

with the corresponding *q-number* gauge transformation rule:

$$(\varphi_1(x) + \varphi_2(x))A'_\mu(x) = -im'(\partial_\mu\Omega(x))\Omega^{-1}(x) + (\varphi_1(x) + \varphi_2(x))\Omega(x)A_\mu(x)\Omega^{-1}(x) \tag{7.13}$$

Using the vacuum state eq. 7.9 we find, for *real-valued* coordinates, eqs. 7.10 – 7.13 become

$$\partial/\partial x_\mu F^a_{\mu\nu} + (g + m'^{-1}\varphi_2)f^{abc}A^{b\mu} F^c_{\mu\nu} = j^a_\nu \tag{7.14}$$

and

$$[i\gamma^\mu(\partial/\partial x^\mu - i(g + m'^{-1}\varphi_2)A_\mu) - m]\psi(x) = 0 \tag{7.15}$$

where

$$F^a_{\mu\nu} = \partial/\partial x^\nu A^a_\mu - \partial/\partial x^\mu A^a_\nu + (g + m'^{-1}\varphi_2)f^{abc}A^b_\mu A^c_\nu \tag{7.16}$$

with the corresponding *q-number* gauge transformation rule:

$$(gm' + \varphi_2(x))A'_\mu(x) = -i\, m'(\partial_\mu\Omega(x))\Omega^{-1}(x) + (gm' + \varphi_2(x))\Omega(x)A_\mu(x)\Omega^{-1}(x) \tag{7.17}$$

7.1.4 C-Number Coupling Constants or Q-Number Coupling Constants?

The above two possible methods for reducing coupling constants to Higgsian vacuum expectation values have different experimental implications. In the case of ElectroWeak gauge fields a c-number coupling constant does not introduce a new interaction with a Higgs particle. In the case of ElectroWeak gauge fields a q-number coupling constant, it does introduce a new interaction with a new Higgs particle. The Higgs particle found at CERN LHC may have been produced from an ElectroWeak gauge field with a q-number coupling constant term.

7.1.5 Strong Interaction Case for Complex-Valued Complexon Coordinates

In the case of the *strong SU(3)* gauge fields the q-number approach would lead to a new interaction of gluons and a Higgs particle corresponding to the field φ_2.

The new dynamic equations for the complexon Yang-Mils field upon replacement of the coupling constant g by a Higgs field using our pseudoQuantization formalism are:

$$D_\mu F^a_{\mu\nu} + (g + m'^{-1}\varphi_2)f^{abc}A^{b\mu} F^c_{\mu\nu} = j^a_\nu \tag{7.18}$$

and

$$[i\gamma^\mu(D_\mu - i(g + m'^{-1}\varphi_2)A_\mu) - m]\psi(x) = 0 \tag{7.19}$$

where

$$F^a_{\mu\nu} = D_\nu A^a_\mu - D_\mu A^a_\nu + (g + m'^{-1}\varphi_2)f^{abc}A^b_\mu A^c_\nu \tag{7.20}$$

where all coordinates are *complex-valued* $x = x_r + ix_i$ with derivatives D_μ given by

$$D_0 = \partial/\partial x^0$$
$$D_k = \partial/\partial x^k + i\,\partial/\partial x_i^k$$

with $x^k = x_r^k$ for k = 1, 2, 3.

The corresponding *q-number* gauge transformation rule:

$$(gm' + \varphi_2(x))A'_\mu(x) = -i\,m'(\partial_\mu\Omega(x))\Omega^{-1}(x) + (gm' + \varphi_2(x))\Omega(x)A_\mu(x)\Omega^{-1}(x) \quad (7.21)$$

Q-number gauge transformations appear in a number of situations. For example, Quantum Electrodynamics has q-number gauge transformations.

7.2 Gravitational Coupling Constant Vacuum Expectation Value Generation

The gravitational coupling constant $g_{CG} = \kappa^{-1} = (4\pi G)^{-1/2}$ appears in the gravitational lagrangian density. An example is the case of an interaction with a Dirac particle:

$$\mathcal{L} = g_{CG}^2\,\sqrt{g}\,R/2 + a\bar{\psi}(i\gamma^\mu\partial/\partial x^\mu - m)\psi \quad (7.22)$$

where a is a coupling constant. and g_{CG} has the dimension of mass. Thus we can introduce a coherent vacuum state

$$|\Phi_G, \Pi_G\rangle = C\exp\{[(2\pi)^3 m/2]^{1/2} g_{CG}[a_2^\dagger(\mathbf{0},m) + a_2(\mathbf{0},m)]\}|0\rangle \quad (7.23)$$

similar to eq. 7.9 that enables us to re-express eq. 7.22 as

$$\mathcal{L} = \varphi_1^2\,\sqrt{g}\,R/2 + a\bar{\psi}(i\gamma^\mu\partial/\partial x^\mu - m)\psi \quad (7.24)$$

or

$$\mathcal{L} = (\varphi_1 + \varphi_2)^2\,\sqrt{g}\,R/2 + c\bar{\psi}(i\gamma^\mu\partial/\partial x^\mu - m)\psi \quad (7.25)$$

using the formalism of section 7.4 with the vacuum state eq. 7.23 throughout.[157] Thus we can directly embody the gravitational constant within our formalism.

If we add the pseudoQuantum fields' lagrangian to the lagrangian of eq. 7.24 we obtain:

$$\mathcal{L} = \varphi_1^2\,\sqrt{g}\,R/2 + a\bar{\psi}(i\gamma^\mu\partial/\partial x^\mu - m)\psi + \partial\varphi_1/\partial x_\mu\partial\varphi_2/\partial x^\mu - m^2\,\varphi_1\varphi_2 \quad (7.26)$$

The dynamic equations for φ_1 and φ_2 are

$$\Box\varphi_2 + m^2\varphi_2 - \varphi_1\,\sqrt{g}\,R = 0 \quad (7.27)$$

and

$$\Box\varphi_1 - m^2\varphi_1 = 0$$

In flat space-time, R = 0 and the equations become free field equations. In curved space-time the curvature scalar term becomes a negative mass counter term reminiscent of the corresponding negative term in the Wheeler-DeWitt equation.

Another possible prototype lagrangian

$$\mathcal{L} = (\varphi_1 + \varphi_2)^2\,\sqrt{g}\,R/2 + a\bar{\psi}(i\gamma^\mu\partial/\partial x^\mu - m)\psi + \partial\varphi_1/\partial x_\mu\partial\varphi_2/\partial x^\mu - m^2\,\varphi_1\varphi_2 \quad (7.28)$$

leads to an interaction between the pseudoQuantum φ_2 field and gravitation:

---

[157] Other variants of these equations are possible such as using the term $g_{CG}\varphi_1\sqrt{g}R/2$ instead of $\varphi_1^2\sqrt{g}R/2$.

$$\Box\varphi_2 + m^2\varphi_2 - (\varphi_1 + \varphi_2)\sqrt{g}\,R = 0 \qquad (7.29)$$
$$\Box\varphi_1 - m^2\varphi_1 = 0$$

7.3 The Eight Coupling Constants and their Eight PseudoQuantum Field Vacuum Expectation Values

As we mentioned in section 7.1 our Grand Unified Theory of Everything (GUTE) has eight coupling constants:

- The Strong interaction coupling constant field g_S.
- The Electromagnetic U(1) coupling constant vacuum expectation value e.
- The ElectroWeak SU(2) coupling constant g_{EW}.
- The ElectroWeak U(1) coupling constant g'_{EW}.
- The Dark ElectroWeak SU(2) coupling constant g_{EWD}.
- The Dark ElectroWeak U(1) coupling constant g'_{EWD}.
- The Layer Group U(4) coupling constant[158] g_V.
- The Generation gauge field U(4) coupling constant g_G.
- The complex gravitational coupling constant $g_{CG} = \kappa^{-1} = (4\pi G)^{-½}$.

Based on the discussions of the previous sections we can define pseudoQuantum fields for these couplings by

- The Electromagnetic U(1) coupling constant vacuum expectation value $\Phi_E = m_E e$.
- The strong interaction coupling constant vacuum expectation value $\Phi_1 = m_1 g_S$.
- The ElectroWeak SU(2) coupling constant vacuum expectation value $\Phi_2 = m_2 g_{EW}$.
- The ElectroWeak U(1) coupling constant vacuum expectation value $\Phi_3 = m_3 g'_{EW}$.
- The Dark ElectroWeak SU(2) coupling constant vacuum expectation value $\Phi_4 = m_4 g_{EWD}$.
- The Dark ElectroWeak U(1) coupling constant vacuum expectation value $\Phi_5 = m_5 g'_{EWD}$
- The Layer Group U(4) coupling constant vacuum expectation value $\Phi_6 = m_6 g_V$.
- The Generation gauge field U(4) coupling constant vacuum expectation value $\Phi_7 = m_7 g_G$.
- The gravitational coupling constant vacuum expectation value $\Phi_8 = g_{CG} = \kappa^{-1} = (4\pi G)^{-½}$.

The eight masses, m_E, m_1, m_2, … , m_7 may be equal or they may have different values. *It is also possible that all masses may be equal to κ^{-1}, which would yield*

- The strong interaction coupling constant vacuum expectation value $\Phi_1 = \kappa^{-1} g_S$.
- The Electromagnetic U(1) coupling constant vacuum expectation value $\Phi_E = \kappa^{-1} e$.
- The ElectroWeak SU(2) coupling constant vacuum expectation value $\Phi_2 = \kappa^{-1} g_{EW}$.
- The ElectroWeak U(1) coupling constant vacuum expectation value $\Phi_3 = \kappa^{-1} g'_{EW}$.
- The Dark ElectroWeak SU(2) coupling constant vacuum expectation value $\Phi_4 = \kappa^{-1} g_{EWD}$.
- The Dark ElectroWeak U(1) coupling constant vacuum expectation value $\Phi_5 = \kappa^{-1} g'_{EWD}$.
- The Layer Group U(4) coupling constant vacuum expectation value $\Phi_6 = \kappa^{-1} g_V$.
- The Generation gauge field U(4) coupling constant vacuum expectation value $\Phi_7 = \kappa^{-1} g_G$.
- The gravitational coupling constant vacuum expectation value $\Phi_8 = g_{CG} = \kappa^{-1} = (4\pi G)^{-½}$.

Then scaling the above vacuum expectation values by κ^{-1} would give:[159]

---

[158] This coupling constant appears in Blaha (2016a).

- The strong interaction coupling constant[160] vacuum expectation value $\Phi_1' = g_S = 1.22$
- The ElectroWeak SU(2) coupling constant vacuum expectation value $\Phi_2' = g_W = 0.619$.
- The Electromagnetic U(1) coupling constant vacuum expectation value $\Phi_E' = 0.303$.
- The ElectroWeak U(1) coupling constant vacuum expectation value $\Phi_3' = g'_{EW} = 0.347$.
- The Dark ElectroWeak SU(2) coupling constant vacuum expectation value $\Phi_4' = g_{EWD}$. (7.30)
- The Dark ElectroWeak U(1) coupling constant vacuum expectation value $\Phi_5' = g'_{EWD}$.
- The Layer Group U(4) coupling vacuum expectation value $\Phi_6 = g_V$.
- The Generation gauge field U(4) coupling constant vacuum expectation value $\Phi_7' = g_G$.
- The gravitational coupling constant vacuum expectation value $\Phi_8' = 1$.

The *scaled* (known) vacuum expectation values,[161] which are mostly in fact the coupling constants, have a comparable range of values[162] as opposed to the range of values for the unscaled constants which range from the ultra-small gravitational vacuum expectation value to values, perhaps, within a few orders of magnitude of unity.

Given the range of known values above, it appears reasonable to conjecture that the unknown values would also be of the order of unity. The known coupling constant values in eq. 7.30 are of comparable value, which suggests that our Theory of Everything, at current energies, may be close to the GUT level at which coupling constants are equal.

End of Extract

---

[159] All coupling constant values are based on data extracted from K. A. Olive et al (Particle Data Group), Chinese Physics **C38**, 090001 (2014).

[160] Based on the running coupling constant value $\alpha_s(M_Z^2) = 0.1193 \pm 0.0016$.

[161] The closeness of all the values to one is suggestive: The value $\alpha = 1$ (or $e = (4\pi)^{1/2} = 3.54$) was the value found in our calculation in the Johnson, Baker, Willey model of QED. Perhaps a larger calculation along the lines of our paper in massless ElectroWeak theory might yield scaled coupling constant values near unity.

[162] The weakness of the ElectroWeak interactions is primarily due to the large masses of the Z and W vector bosons – not the values of their coupling constants g and g'.

REFERENCES

Bjorken, J. D., Drell, S. D., 1964, *Relativistic Quantum Mechanics* (McGraw-Hill, New York, 1965).

Bjorken, J. D., Drell, S. D., 1965, *Relativistic Quantum Fields* (McGraw-Hill, New York, 1965).

Blaha, S., 1998, *Cosmos and Consciousness* (Pingree-Hill Publishing, Auburn, NH, 1998).

_______, 2002, *A Finite Unified Quantum Field Theory of the Elementary Particle Standard Model and Quantum Gravity Based on New Quantum Dimensions™ & a New Paradigm in the Calculus of Variations* (Pingree-Hill Publishing, Auburn, NH, 2002).

_______, 2003, *A Finite Unified Quantum Field Theory of the Elementary Particle Standard Model and Quantum Gravity Based on New Quantum Dimensions™ and a New Paradigm in the Calculus of Variations* (Pingree-Hill Publishing, Auburn, NH, 2003).

_______, 2004, *Quantum Big Bang Cosmology: Complex Space-time General Relativity, Quantum Coordinates,™ Dodecahedral Universe, Inflation, and New Spin 0, ½, 1 & 2 Tachyons & Imagyons* (Pingree-Hill Publishing, Auburn, NH, 2004).

______, *2005a, Quantum Theory of the Third Kind: A New Type of Divergence-free Quantum Field Theory Supporting a Unified Standard Model of Elementary Particles and Quantum Gravity based on a New Method in the Calculus of Variations* (Pingree-Hill Publishing, Auburn, NH, 2005).

______, 2005b, *The Metatheory of Physics Theories, and the Theory of Everything as a Quantum Computer Language* (Pingree-Hill Publishing, Auburn, NH, 2005).

______, 2005c, *The Equivalence of Elementary Particle Theories and Computer Languages: Quantum Computers, Turing Machines, Standard Model, Superstring Theory, and a Proof that Gödel's Theorem Implies Nature Must Be Quantum* (Pingree-Hill Publishing, Auburn, NH, 2005).

______, 2006a, *The Foundation of the Forces of Nature* (Pingree-Hill Publishing, Auburn, NH, 2006).

______, 2006b, *A Derivation of ElectroWeak Theory based on an Extension of Special Relativity; Black Hole Tachyons; & Tachyons of Any Spin.* (Pingree-Hill Publishing, Auburn, NH, 2006).

_______, 2007a, *Physics Beyond the Light Barrier: The Source of Parity Violation, Tachyons, and A Derivation of Standard Model Features* (Pingree-Hill Publishing, Auburn, NH, 2007).

______, 2007b, *The Origin of the Standard Model: The Genesis of Four Quark and Lepton Species, Parity Violation, the ElectroWeak Sector, Color SU(3), Three Visible Generations of Fermions, and One Generation of Dark Matter with Dark Energy* (Pingree-Hill Publishing, Auburn, NH, 2007).

______*, 2008a, A Direct Derivation of the Form of the Standard Model From GL(16) (Pingree-Hill Publishing, Auburn, NH, 2008).*

______, 2008b, *A Complete Derivation of the Form of the Standard Model With a New Method to Generate Particle Masses Second Edition* (Pingree-Hill Publishing, Auburn, NH, 2008)

______, 2009, *The Algebra of Thought & Reality: The Mathematical Basis for Plato's Theory of Ideas, and Reality Extended to Include A Priori Observers and Space-Time Second Edition* (Pingree-Hill Publishing, Auburn, NH, 2009).

______, 2010a, *Operator Metaphysics: A New Metaphysics Based on a New Operator Logic and a New Quantum Operator Logic that Lead to a Mathematical Basis for Plato's Theory of Ideas and Reality* (Pingree-Hill Publishing, Auburn, NH, 2010).

______, 2010b, *The Standard Model's Form Derived from Operator Logic, Superluminal Transformations and GL(16)* (Pingree-Hill Publishing, Auburn, NH, 2010).

______, 2011a, *21st Century Natural Philosophy Of Ultimate Physical Reality* (McMann-Fisher Publishing, Auburn, NH, 2011).

______, 2011b, *All the Universe! Faster Than Light Tachyon Quark Starships & Particle Accelerators with the LHC as a Prototype Starship Drive Scientific Edition* (Pingree-Hill Publishing, Auburn, NH, 2011).

______, 2011c, *From Asynchronous Logic to The Standard Model to Superflight to the Stars* (Blaha Research, Auburn, NH, 2011).

______, 2012a, *From Asynchronous Logic to The Standard Model to Superflight to the Stars volume 2: Superluminal CP and CPT, U(4) Complex General Relativity and The Standard*

Model, Complex Vierbein General Relativity, Kinetic Theory, Thermodynamics (Blaha Research, Auburn, NH, 2012).

______, 2012b, *Standard Model Symmetries, And Four And Sixteen Dimension Complex Relativity; The Origin Of Higgs Mass Terms* (Blaha Reasearch, Auburn, NH, 2012).

______, 2013a, *Multi-Stage Space Guns, Micro-Pulse Nuclear Rockets, and Faster-Than-Light Quark-Gluon Ion Drive Starships* (Blaha Research, Auburn, NH, 2013).

______, 2013b, *The Bridge to Dark Matter; A New Sister Universe; Dark Energy; Inflatons; Quantum Big Bang; Superluminal Physics; An Extended Standard Model Based on Geometry* (Blaha Reasearch, Auburn, NH, 2013).

______, 2014a, *Universes and Multiverses: From a New Standard Model to a Physical Multiverse; The Big Bang; Our Sister Universe's Wormhole; Origin of the Cosmological Constant, Spatial Asymmetry of the Universe, and its Web of Galaxies; A Baryonic Field between Universes and Particles; Flatverse Extended Wheeler-DeWitt Equation* (Blaha Reasearch, Auburn, NH, 2014).

______, 2014b, *All the Multiverse! Starships Exploring the Endless Universes of the Cosmos Using the Baryonic Force* (Blaha Research, Auburn, NH, 2014).

______, 2014c, *All the Multiverse! II Between Multiverse Universes: Quantum Entanglement Explained by the Multiverse Coherent Baryonic Radiation Devices – PHASERs Neutron Star Multiverse Slingshot Dynamics Spiritual and UFO Events, and the Multiverse Microscopic Entry into the Multiverse* (Blaha Research, Auburn, NH, 2014).

______, 2015a, *PHYSICS IS LOGIC PAINTED ON THE VOID: Origin of Bare Masses and The Standard Model in Logic, U(4) Origin of the Generations, Normal and Dark Baryonic Forces, Dark Matter, Dark Energy, The Big Bang, Complex General Relativity, A Megaverse of Universe Particles* (Blaha Research, Auburn, NH, 2015).

______, 2015b, *PHYSICS IS LOGIC Part II: The Theory of Everything, The Megaverse Theory of Everything, U(4)⊗U(4) Grand Unified Theory (GUT), Inertial Mass = Gravitational Mass, Unified Extended Standard Model and a New Complex General Relativity with Higgs Particles, Generation Group Higgs Particles* (Blaha Research, Auburn, NH, 2015).

______, 2015c, *The Origin of Higgs ("God") Particles and the Higgs Mechanism: Physics is Logic III, Beyond Higgs – A Revamped Theory With a Local Arrow of Time, The Theory of Everything Enhanced, Why Inertial Frames are Special, Universes of the Mind* (Blaha Research, Auburn, NH, 2015).

______, 2015d, *The Origin of the Eight Coupling Constants of The Theory of Everything: U(8) Grand Unified Theory of Everything (GUTE), S^8 Coupling Constant Symmetry, Space-Time Dependent Coupling Constants, Big Bang Vacuum Coupling Constants, Physics is Logic IV* (Blaha Research, Auburn, NH, 2015).

______, 2016a, *New Types of Dark Matter, Big Bang Equipartition, and A New U(4) Symmetry in the Theory of Everything: Equipartition Principle for Fermions, Matter is 83.33% Dark, Penetrating the Veil of the Big Bang, Explicit QFT Quark Confinement and Charmonium, Physics is Logic V* (Blaha Research, Auburn, NH, 2016).

______, 2016b, *The Periodic Table of the 192 Quarks and Leptons in The Theory of Everything: The U(4) Layer Group, Physics is Logic VI* (Blaha Research, Auburn, NH, 2016).

______, 2016c, *New Boson Quantum Field Theory, Dark Matter Dynamics, Dark Matter Fermion Layer Mixing, Genesis of Higgs Particles, New Layer Higgs Masses, Higgs Coupling Constants, Non-Abelian Higgs Gauge Fields, Physics is Logic VII* (Blaha Research, Auburn, NH, 2016).

______, 2016d, *Unification of the Strong Interactions and Gravitation: Quark Confinement Linked to Modified Short-Distance Gravity; Physics is Logic VIII* (Blaha Research, Auburn, NH, 2016).

______, 2016e, *MoND: Unification of the Strong Interactions and Gravitation II, Quark Confinement Linked to Large-Scale Gravity, Physics is Logic IX* (Blaha Research, Auburn, NH, 2016).

______, 2016f, *CQMechanics: A Unification of Quantum & Classical Mechanics, Quantum/Semi-Classical Entanglement, Quantum/Classical Path Integrals, Quantum/Classical Chaos* (Blaha Research, Auburn, NH, 2016).

______, 2016g, *GEMS: Unified Gravity, ElectroMagnetic and Strong Interactions: Manifest Quark Confinement, A Solution for the Proton Spin Puzzle, Modified Gravity on the Galactic Scale* (Pingree Hill Publishing, Auburn, NH, 2016).

Chrystal, G., 1961, *Textbook of Algebra Part One* (Dover Publications, Inc., New York, 1961).

Eddington, A. S., 1952, *The Mathematical Theory of Relativity* (Cambridge University Press, Cambridge, U.K., 1952).

Fant, Karl M., 2005, *Logically Determined Design: Clockless System Design With NULL Convention Logic* (John Wiley and Sons, Hoboken, NJ, 2005).
Heitler, W., 1954, *The Quantum Theory of Radiation* (Claendon Press, Oxford, UK, 1954).

Huang, Kerson, 1992, *Quarks, Leptons & Gauge Fields 2nd Edition* (World Scientific Publishing Company, Singapore, 1992).

Misner, C. W., Thorne, K. S., and Wheeler, J. A., 1973, *Gravitation* (W. H. Freeman, New York, 1973).

Sagan, H., 1993, *Introduction to the Calculus of Variations* (Dover Publications, Mineola, NY, 1993).

Sakurai, J. J., 1964, *Invariance Principles and Elementary Particles* (Princeton University Press, Princeton, NJ, 1964).

Streater, R. F. and Wightman, A. S., 2000, *PCT, Spin, Statistics, and All That* (Princeton University Press, Princeton, NJ 2000).

Weinberg, S., 1972, *Gravitation and Cosmology* (John Wiley and Sons, New York, 1972).

Weinberg, S., 1995, *The Quantum Theory of Fields Volume I* (Cambridge University Press, New York, 1995).

Weyl, H., 1950, *Space, Time, Matter* (Dover, New York, 1950).

Weyl, H., (Tr. S. Pollard et al), 1987, *The Continuum* (Dover Publications, New York, 1987).

INDEX

About the Author

Stephen Blaha is a well known Physicist and Man of Letters with interests in Science, Society and civilization, the Arts, and Technology. He had an Alfred P. Sloan Foundation scholarship in college. He received his Ph.D. in Physics from Rockefeller University. He has served on the faculties of several major universities. He was also a Member of the Technical Staff at Bell Laboratories, a manager at the Boston Globe Newspaper, a Director at Wang Laboratories, and President of Blaha Software Inc and of Janus Associates Inc. (NH).

Among other achievements he was a co-discoverer of the "r potential" for heavy quark binding developing the first (and still the only demonstrable) non-abelian gauge theory with an "r" potential; first suggested the existence of topological structures in superfluid He-3; first proposed Yang-Mills theories would appear in condensed matter phenomena with non-scalar order parameters; first developed a grammar-based formalism for quantum computers and applied it to elementary particle theories; first developed a new form of quantum field theory without divergences (thus solving a major 60 year old problem that enabled a unified theory of the Standard Model and Quantum Gravity without divergences to be developed); first developed a formulation of complex General Relativity based on analytic continuation from real space-time; first developed a generalized non-homogeneous Robertson-Walker metric that enabled a quantum theory of the Big Bang to be developed without singularities at $t = 0$; first generalized Cauchy's theorem and Gauss' theorem to complex, curved multi-dimensional spaces; received Honorable Mention in the Gravity Research Foundation Essay Competition in 1978; first developed a physically acceptable theory of faster-than-light particles; first derived a composition of extrema method in the Calculus of Variations; first quantitatively suggested that inflationary periods in the history of the universe were not needed; first proved Gödel's Theorem implies Nature must be quantum; provided a new alternative to the Higgs Mechanism, and Higgs particles, to generate masses; first showed how to resolve logical paradoxes including Gödel's Undecidability Theorem by developing Operator Logic and Quantum Operator Logic; first developed a quantitative harmonic oscillator-like model of the life cycle, and interactions, of civilizations; first showed how equations describing superorganisms also apply to civilizations. A recent book shows his theory applies successfully to the past 14 years of history and to *new* archaeological data on Andean and Mayan civilizations as well as Early Anatolian and Egyptian civilizations.

He first developed an axiomatic derivation of the forms of The Standard Model from geometry – space-time properties – The Extended Standard Model. It has a Dark Matter sector that approximates the ElectroWeak sector with Dark doublets and Dark gauge interactions. It also uses quantum coordinates to remove infinities that crop up in most interacting quantum field theories and additionally to remove the infinities that appear in the Big Bang and generate an inflationary growth of the universe. The Extended Standard Model has an ultra-high energy GUT (Grand Unified Theory) limit with a $U(4)\otimes U(4)$ symmetry; and can be united with gravitation to form a Theory of Everything. (See *Physics is Logic Part II.*)

Blaha has had a major impact on a succession of elementary particle theories: his Ph.D. thesis (1970), and papers, showed that quantum field theory calculations to all orders in ladder approximations could not give scaling deep inelastic electron-nucleon scattering. He later showed the eigenvalue equation for the fine structure constant α in Johnson-Baker-Willey QED had a zero at $\alpha = 1$ not 1/137 by solving the Schwinger-Dyson equations to all orders in an approximation that agreed with exact results to 4th order in α thus ending interest in this theory. In 1979 at Prof. Ken Johnson's (MIT) suggestion he calculated the proton-neutron mass difference in the MIT bag model and found the result had the wrong sign reducing interest in the bag model. These results all appear in Physical Review papers. In the 2000's he repeatedly pointed out the shortcomings of SuperString theory and showed that The Standard Model's form could be derived from space-time geometry by an extension of Lorentz transformations to faster than light transformations. This deeper space-time basis greatly increases the possibility that it is part of THE fundamental theory.Recently, Blaha showed that the Weak interactions differed significantly from the Strong, electromagnetic and gravitation interactions in important respects while these interactions had similar features, and suggested that ElectroWeak theory, which is essentially a glued union of the Weak interactions and Electromagnetism, possibly modulo unknown Higgs particle features, be replaced by a unified theory of the other interactions combined with a stand-alone Weak interaction theory. Blaha also showed that, if Charmonium calculations are taken seriously, the Strong interaction coupling constant is only a factor of five larger than the electromagnetic coupling constant, and thus Strong interaction perturbation theory would make sense and yield physically meaningful results.

In graduate school (1965-71) he wrote substantial papers in elementary particles and group theory: The Inelastic E- P Structure Functions in a Gluon Model. Phys. Lett. B40:501-502,1972; Deep-Inelastic E-P Structure Functions In A Ladder Model With Spin 1/2 Nucleons, Phys.Rev. D3:510-523,1971; Continuum Contributions To The Pion Radius, Phys. Rev. 178:2167-2169,1969; Character Analysis of U(N) and SU(N), J. Math. Phys. 10, 2156 (1969); and The Calculation of the Irreducible Characters of the Symmetric Group in Terms of the Compound Characters, (Published as Blaha's Lemma in D. E. Knuth's book: *The Art of Computer Programming Vols. 1 – 4*).

In the early 1980's Blaha was also a pioneer in the development of UNIX for financial, scientific and Internet applications: benchmarked UNIX versions showing that block size was critical for UNIX performance, developing financial modeling software, starting database benchmarking comparison studies, developing Internet-like UNIX networking (1982) and developing a hybrid shell programming technique (1982) that was a precursor to the PERL programming language. He was also the manager of the AT&T ten-year future products development database. His work helped lead to commercial UNIX on computers such as Sun Micros, IBM AIX minis, and Apple computers.

In the 1980's he pioneered the development of PC Desktop Publishing on laser printers. and was nominated for three "Awards for Technical Excellence" in 1987 by PC Magazine for PC software products that he designed and developed.

Recently he has developed a theory of Megaverses – actual universes of which our universe is one – with quantum particle-like properties based on the Wheeler-DeWitt equation of Quantum Gravity. He has developed a theory of a baryonic force, which had been conjectured many years ago, and estimated the strength of the force based on discrepancies in measurements of the gravitational constant G. This force, operative in 15-dimensinal space, can be used to escape from our universe in "uniships" which are the equivalent of the faster-than-light starships proposed in the author's earlier books. Thus travel to other universes, as well as to other stars is possible.

Blaha also considered the complexified Wheeler-DeWitt equation and showed that its limitation to real-valued coordinates and metrics generated a Cosmological Constant in the Einstein equations.

The author has also recently written a series of books on the serious problems of the United States and their solution as well as a book on the decline of Mankind that will follow from current social and genetic trends in Mankind.

In the past twelve years Dr. Blaha has written over 40 books on a wide range of topics. Some recent major works are: *From Asynchronous Logic to The Standard Model to Superflight to the Stars, All the Universe!, SuperCivilizations: Civilizations as Superorganisms, America's Future: an Islamic Surge, ISIS, al Qaeda, World Epidemics, Ukraine, Russia-China Pact, US Leadership Crisis,The Rises and Falls of Man – Destiny – 3000 AD: New Support for a Superorganism MACRO-THEORY of CIVILIZATIONS From CURRENT WORLD TRENDS and NEW Peruvian, Pre-Mayan, Mayan, Anatolian, and Early Egyptian Data, with a Projection to 3000 AD,* and *Mankind in Decline: Genetic Disasters, Human-Animal Hybrids, Overpopulation, Pollution, Global Warming, Food and Water Shortages, Desertification, Poverty, Rising Violence, Genocide, Epidemics, Wars, Leadership Failure.*

He has taught approximately 4,000 students in undergraduate, graduate, and postgraduate corporate education courses primarily in major universities, and large companies and government agencies.

The above paragraphs summarize much of his work over the past fifty years. This work is fully documented. He continues to engage in research and writing at Blaha Research.

www.ingramcontent.com/pod-product-compliance
Lightning Source LLC
Chambersburg PA
CBHW082007190326
41458CB00010B/3108
* 9 7 8 0 9 9 8 4 5 7 5 0 5 *